Wireless Communications Under Hostile Jamming: Security and Efficiency

Tongtong Li • Tianlong Song • Yuan Liang

Wireless Communications Under Hostile Jamming: Security and Efficiency

Tongtong Li
Department of Electrical
and Computer Engineering
Michigan State University
East Lansing, MI, USA

Tianlong Song
Zillow Inc
Seattle, WA, USA

Yuan Liang
Department of Electrical
and Computer Engineering
Michigan State University
East Lansing, MI, USA

ISBN 978-981-13-4509-8 ISBN 978-981-13-0821-5 (eBook)
https://doi.org/10.1007/978-981-13-0821-5

Softcover re-print of the Hardcover 1st edition 2018

This Springer imprint is published by the registered company Springer Nature Singapore Pte Ltd.
The registered company address is: 152 Beach Road, #21-01/04 Gateway East, Singapore 189721, Singapore

"When a door closes, a window opens."

To the designers of next generation secure wireless communication systems

Preface

As we are entering the era of Internet of Things (IoT), the importance of secure, efficient, and reliable wireless communications cannot be overemphasized. It has become the key enabler in information exchange, remote sensing, monitoring and control, and is finding applications in nearly every field.

Mainly due to limited total available spectrum and lack of a protective physical boundary, wireless communication is facing much more serious challenges in security and capacity than its wirelined counterpart. In addition to the time and frequency dispersions, wireless signals are subjected to unauthorized detection, interception/eavesdropping, and hostile jamming.

In hostile jamming, the jammer intends to disable the legitimate transmission by saturating the receiver with noise or false information through deliberate radiation of radio signals. Comparing to passive attacks such as unauthorized detection and eavesdropping, hostile jamming is an active attack and is much more destructive. It is one of the most commonly used techniques for limiting the effectiveness of an opponent's communication. Along with the wide spread of advanced wireless devices, especially with the advent of user configurable intelligent devices, jamming attack is no longer limited to battlefield or military-related events, but has become an urgent and serious threat to civilian communications as well.

This book is targeted at secure and efficient anti-jamming system design for civilian applications. In communication system design, security is often achieved with a sacrifice on efficiency. Unlike in military applications where rich spectral diversity is generally available, civilian applications are often highly limited in both bandwidth and power. As can be seen, the biggest challenge here lies in how to design secure anti-jamming systems which are robust and reliable and yet possess high spectral and power efficiency. For practical applications, the anti-jamming systems need to be feasible and cost effective, and ideally, can be implemented through smooth upgrades of existing systems. These are the major criteria and guidelines in our study and research on anti-jamming system design.

In this book, we revisit existing jamming patterns and introduce new jamming patterns. For a long time, it was believed that only strong jamming is really harmful. It turns out that this is only partially true. If the authorized signal coding space and pulse shaping filter are observed by the jammer, the jammer can then mimic the characteristics of the authorized signal and launch disguised jamming, which is of the same power level as that of the authorized signal, such that the receiver cannot distinguish the jamming from the authorized signal. The symmetricity between the jamming and the authorized signal leads to complete communication failure. To make things worse, disguised jamming attack happens at the symbol level and is immune to bit-level channel coding. That is, reducing the code rate at the bit level will not improve the system performance under disguised jamming. Motivated by this observation, a considerable amount of contents in this book are focused on secure and efficient wireless system design under disguised jamming.

We analyze the limitations of existing communication systems and anti-jamming techniques. We point out that existing communication systems do not possess sufficient security features and are generally very fragile under disguised jamming. Most communication systems, such as EDGE and OFDM, are designed for efficient and accurate information transmission from the source to destination, and do not have inherent security features. When attacked by disguised jamming, the deterministic capacity of these systems will be reduced to zero. Existing work on anti-jamming system design or jamming mitigation is mainly based on spread spectrum techniques, including CDMA and frequency hopping (FH). Both CDMA and FH systems possess anti-jamming features by exploiting frequency diversity over large spectrum. However, while these systems work reasonably well for voice-centric communication, their security feature and information capacity are far from adequate and acceptable for today's high-speed multimedia wireless services.

We present a whole family of innovative and feasible anti-jamming techniques, which can strengthen the inherent security of the 3G, 4G, and the upcoming 5G systems with minimal and inexpensive changes to the existing CDMA, frequency hopping, TDMA, and OFDM schemes. We enhance the security of CDMA systems through secure scrambling: in addition to its robustness to narrow-band jamming, CDMA with secure scrambling is proved to be secure and efficient under disguised jamming. We develop innovative frequency hopping techniques, known as message-driven frequency hopping (MDFH) and collision-free frequency hopping (CFFH). In MDFH, part of the encrypted message stream acts as the PN sequence and transmitted through hopping frequency control. Transmission through hopping frequency control essentially introduces another dimension to the signal space, and the corresponding coding gain can increase the system efficiency by multiple times. On the other hand, CFFH is developed to resolve the self-collision problem in conventional frequency hopping. CFFH can ensure that each user hops in a pseudo-random manner, and different users always transmit on nonoverlapping sets of subcarriers. CFFH can be directly applied to any multiband multiaccess systems to achieve secure and dynamic spectral access control. CFFH is proved to be particularly effective in mitigating partial band jamming and follower jamming. When applied to OFDMA system, the resulted CFFH-OFDM has the same spectral efficiency as the original OFDM system. We develop an innovative secure precoding scheme which can achieve constellation randomization, break the symmetricity between the authorized signal and jamming, and hence achieve secure and efficient transmission under disguised jamming. We apply it to the OFDM system and come up with the securely precoded OFDM (SP-OFDM). While achieving strong resistance against disguised jamming, SP-OFDM has the same high spectral efficiency as the traditional OFDM system.

These new anti-jamming techniques are the main contents and major contributions of this book, and they are presented in Chaps. 2, 3, 4, and 5. Along with detailed descriptions on system design, we also provide benchmarks for system performance evaluation under various jamming scenarios, through theoretical capacity analysis as well as numerical examples. Overall, the most important and exciting finding here is: secure yet efficient anti-jamming wireless communications can be achieved by integrating modern cryptographic techniques into physical layer transceiver design.

Moreover, in this book, we examine the ultimate game between the legitimate user and jammer, where a power-limited authorized user and a power-limited jammer are operating swiftly and independently over the same spectrum consisting of multiple bands. We also look into the future research directions and share our thoughts on jamming and anti-jamming in 5G IOT systems and 5G wireless network design and performance evaluation under hostile environments.

Embodied with design principles, theoretical analysis, and practical design examples, we sincerely wish that this book can serve as a good reference for all of those who are designers and would-be designers of secure and efficient wireless communication systems under intentional interference, including both academic researchers and professionals in industry.

East Lansing, MI, USA Tongtong Li
Seattle, WA, USA Tianlong Song
East Lansing, MI, USA Yuan Liang
September 14, 2018

Acknowledgments

This book is rooted in the research on jamming, anti-jamming, and physical layer security of wireless systems at the Broadband Access Wireless Communication (BAWC) Lab of Michigan State University from 2002 to 2018. We are deeply thankful to our previous lab members and co-workers who made direct or indirect contributions to this work: Dr. Weiguo Liang, Dr. Huahui Wang, Dr. Qi Ling, Dr. Leonard Lightfoot, Dr. Lei Zhang, Dr. Mai Abdelhakim, Dr. Ahmed Alahmadi, Dr. Zhaoxi Fang, Dr. Run Tian, Dr. Zhe Wang, and Mr. Yu Zheng, who is working toward his Ph.D. degree. We would like to thank colleagues from Springer for inviting us to write this monograph and for their help during the book preparation and publication process.

We are deeply thankful to all who have loved, supported, and helped us. Tongtong wishes to thank her previous graduate advisors Prof. Xinyi Chang, Prof. Donggao Deng, Prof. Zhi Ding, and Prof. Jitendra Tugnait for their mentoring and help and thank her parents Mr. Jicang Li and Ms. Zhiming Zhang for planting the seed of love for science in her heart when she was young. She wants to thank her husband, Prof. Jian Ren, who happens to work on cryptography and network security, for his help on cryptographic technique selection and security analysis and his unconditional support over the years, and also their kids, Brighty and Ellen, for their love and the invaluable happiness they brought to her life.

Tianlong would like to thank his advisors all the way along his journey to become a professional in this field, Prof. Qing Chang, Prof. Suqin He, and Prof. Tongtong Li, and also thank his wife, Dr. Jing Yu, and his parents, Mr. Zhuangshan Song and Ms. Lanqin Ji, for their support and understanding throughout the years, when he spent the time on his research and pursued his dreams.

Yuan would like to thank his current and previous academic advisors, Prof. Tongtong Li, Prof. Lenan Wu, and Prof. Shulin Zhang, for their guidance and help and also thank his late father, Mr. Jianmin Liang, and his mother especially, Ms. Li Jiang, for the dedication and support.

This work is partially supported by National Science Foundation under awards CNS-0746811 and CNS-1217206 and ECCS-1744604. We sincerely appreciate the generous support from NSF and the research community. Without that, our research would not have been possible.

Contents

Acronyms

3G	Third generation
3GPP	Third Generation Partnership Project
4G	Fourth generation
5G	Fifth generation
AES	Advanced encryption standard
AJ-MDFH	Anti-jamming message-driven frequency hopping
AM	Amplitude modulation
AVC	Arbitrarily varying channel
AWGN	Additive white Gaussian noise
BER	Bit error rate
BP	Belief propagation
BPF	Bandpass filter
BPSK	Binary phase shift keying
CDF	Cumulative distribution function
CDMA	Code division multiple access
CF	Characteristic function
CFFH	Collision-free frequency hopping
CF-MDFH	Collision-free message-driven frequency hopping
CLT	Central limit theorem
CP	Cyclic prefix
CSCG	Circularly symmetric complex Gaussian
CSI	Channel state information
DES	Data encryption standard
DFT	Discrete Fourier transform
DS	Direct sequence
DVB-S.2	Digital video broadcasting-satellite-second generation
E-MDFH	Enhanced message-driven frequency hopping
EDGE	Enhanced Data rates for GSM Evolution
ETSI	European Telecommunications Standards Institute
FFT	Fast Fourier transform
FH	Frequency hopping

FHMA	Frequency hopping multiple access
FM	Frequency modulation
FSK	Frequency-shift keying
GPS	Global positioning system
GSM	Global System for Mobile communications
HDR	High data rate
HF	High frequency
i.i.d.	Independent and identically distributed
IA	Interference avoidance
IDFT	Inverse discrete Fourier transform
IEEE	The Institute of Electrical and Electronic Engineers
IFFT	Inverse fast Fourier transform
IoT	Internet of Things
ISI	Intersymbol interference
ITU	International Telecommunication Union
JSI	Jammer state information
JSR	Jamming to signal ratio
KKT	Karush-Kuhn-Tucker
LDPC	Low density parity check
LFSR	Linear feedback shift register
LOS	Line of sight
LTE	Long-Term Evolution
LMS	Least mean square
MAI	Multiuser access interference
MAP	Maximum a posteriori probability
MC-AJ-MDFH	Multi-carrier anti-jamming message-driven frequency hopping
MC-CDMA	Multi-carrier code division multiple access
MDFH	Message-driven frequency hopping
MI	Mutual information
MIMO	Multiple-input and multiple-output
ML	Maximum likelihood
MMSE	Minimum mean square error
NN	Nearest neighbor
OFDM	Orthogonal frequency division multiplexing
OFDMA	Orthogonal frequency division multiple access
OSTC	Orthogonal space time codes
PAM	Pulse amplitude modulation
PAPR	Peak to average power ratio
PDF	Probability density function
PM	Phase modulation
PN	Pseudo noise
PSD	Power spectral density
PSK	Phase shift keying
QAM	Quadrature amplitude modulation
QPSK	Quadrature phase shift keying

RMS	Root mean square
RV	Random variable
SINR	Signal-to-interference-and-noise ratio
SJNR	Signal-to-jamming-and-noise ratio
SJR	Signal-to-jamming ratio
SNR	Signal-to-noise ratio
SP-OFDM	Securely precoded orthogonal frequency division multiplexing
SS	Spread spectrum
STBC	Space time block code
STC	Space time coding
TDMA	Time division multiple access
UMTS	Universal Mobile Telecommunications System
WSS	Wide sense stationary
WSSUS	Wide sense stationary uncorrelated scattering

Chapter 1
Introduction

1.1 Hostile Jamming: A Brief Introduction

Mainly due to limited total available spectrum and lack of a protective physical boundary, wireless communication is facing much more serious challenges in security and capacity than its wirelined counterpart. In addition to the time and frequency dispersions, wireless signals are subjected to unauthorized detection, interception/eavesdropping, and hostile jamming.

In hostile jamming, the jammer intends to disable the legitimate transmission by saturating the receiver with noise or false information through deliberate radiation of radio signals [1–3]. Comparing to passive attacks such as unauthorized detection and eavesdropping, hostile jamming is an active attack and much more destructive. It is one of the most commonly used techniques for limiting the effectiveness of an opponent's communication. Along with the wide spread of advanced wireless devices, especially with the advent of user-configurable intelligent devices, jamming attack is no longer limited to battlefield or military-related events, but has become an urgent and serious threat to civilian communications as well. For example, the primary user emulation attack in cognitive radios, in which the malicious user mimics the characteristics of the primary user's signals to prevent the secondary users from effective transmission, is essentially a jamming attack [4]. Moreover, as is well known, jamming has been a serious issue in the Global Positioning System (GPS) system [5] and in the remote control of unmanned aerial vehicles.

1.1.1 Existing Jamming Models

We first revisit the existing jamming models. In literature, jamming signals are generally categorized into three classes: (i) *Band jamming*, generally modeled as a zero-mean wide-sense stationary Gaussian random process with a flat power

T. Li et al., *Wireless Communications Under Hostile Jamming: Security and Efficiency*, https://doi.org/10.1007/978-981-13-0821-5_1

spectral density (PSD) over the bandwidth of interest. Band jamming is further classified into *full-band jamming* [1] and *partial-band jamming* [2]; (ii) *tone jamming*, typically a sinusoid waveform whose power is concentrated on the carrier frequency. Tone jamming includes single-tone jamming and multi-tone jamming [3]. (iii) *Partial-time jamming*, modeled as a two-state Markov process, for which the jammer is on in state 1 and is off in state 0; state 1 occurs with probability ρ, and state 0 occurs with probability $(1-\rho)$ [6].

Existing work on jamming detection and jamming prevention was generally targeted at a particular jamming model at a time. That is, the jamming pattern is assumed to be known and invariant during the signal transmission period (e.g., see [7, 8]). In practice, however, the jammer may very likely switch frequently from one pattern to another, with each jamming pattern only lasting a very short period of time. In other words, the jammer may launch smart, cognitive jamming, also known as adaptive jamming or time-varying jamming [9].

Let $J(t)$ represent the hostile jamming interference, which can either be stationary or nonstationary. Let $R_J(t,\tau) = E\{J(t+\tau)J(t)\}$ be the autocorrelation function of $J(t)$. The time-varying power spectral density is defined as

$$S_J(t,f) = \mathcal{F}\{R_J(t,\tau)\} = \int_{-\infty}^{\infty} R_J(t,\tau)e^{-j2\pi f\tau}d\tau. \tag{1.1}$$

The time-varying jamming power is given by $P_J(t) = \int_{-\infty}^{\infty} S_J(t,f)df$. We assume that $P_J(t)$ is finite and $P_J(t) \leq P_{J,max}$, where $P_{J,max}$ is the maximum jamming power. The reason that we allow the jamming power to be time-varying, rather than always using the total available jamming power, is because that strong jamming that uses the whole jamming power may not always be the worst jamming [10]. If $J(t)$ is wide-sense stationary, then $R_J(t,\tau)$, $S_J(t,f)$ and $P_J(t)$ all become time-invariant.

It turns out that *all the existing jamming models can be characterized using the time-varying power spectral density,* $S_J(t,f)$. In fact, let f_0 and f_1 be the start and ending frequency of the available frequency band, respectively, $[t_0,t_1]$ the time duration period of the message signal, and P_J the constant jamming power.

- If $S_J(t,f) = \frac{P_J}{f_1-f_0} \triangleq N_J, \forall f \in [f_0,f_1], \forall t \in [t_0,t_1]$, then $J(t)$ with PSD $S_J(t,f)$ is reduced to the traditional full-band jamming. Partial-band jamming can be defined in a similar manner.
- If $S_J(t,f) = P_J\delta(f-f_k), \forall t \in [t_0,t_1]$, where $f_k \in [f_0,f_1]$ and δ is the Dirac delta function, then we obtain the single-tone jamming. For multi-tone jamming,

$$S_J(t,f) = \sum_{k=1}^{K} P_J(t,k)\delta(f-f_k), \ s.t. \sum_{k=1}^{K} P_J(t,k) = P_J, \tag{1.2}$$

where K is the number of jamming tones and $P_J(t,k)$ stands for the jamming power allocated for the kth tone f_k.

- If $\forall f \in [f_0, f_1]$,

$$S_J(t, f) = \begin{cases} 0, & \text{if } t \in [t_0, t_m) \text{ or } t \in (t_n, t_1], \\ \frac{P_J}{f_1 - f_0}, & \text{if } t \in [t_m, t_n], \end{cases} \tag{1.3}$$

where t_m and t_n ($t_m \leq t_n$) are certain intermediate time instants within $[t_0, t_1]$, then we obtain the partial-time jamming.

1.1.2 Disguised Jamming: A New Concept

In [11, 12], we introduced the concept of *disguised jamming*. *Disguised jamming* denotes the case where the jamming is highly correlated with the authorized signal and has a power level close or equal to the signal power. More specifically, let $s(t)$ and $J(t)$ be the authorized signal and jamming interference, respectively. Define

$$\rho = \frac{1}{T_0\sqrt{P_s P_J}} \int_{t_1}^{t_2} s(t)J^*(t)dt \tag{1.4}$$

as the normalized cross-correlation coefficient of $s(t)$ and $J(t)$ over the time period $[t_1, t_2]$, where $T_0 = t_2 - t_1$, $P_s = \frac{1}{T_0}\int_{t_1}^{t_2} |s(t)|^2 dt$ and $P_J = \frac{1}{T_0}\int_{t_1}^{t_2} |J(t)|^2 dt$. We say that $J(t)$ is a disguised jamming to signal $s(t)$ over $[t_1, t_2]$ if:

1. $J(t)$ and $s(t)$ are highly correlated. More specifically, $|\rho| > \rho_0$, where ρ_0 is an application-oriented, predefined correlation threshold.
2. The jamming-to-signal ratio (JSR) is close to 0 dB. More specifically, $|\frac{P_J}{P_s} - 1| < \epsilon_P$, where ϵ_P is an application-oriented, predefined jamming-to-signal ratio threshold.

When the constellation Ω and the pulse shaping filter of the information signal are known to the jammer, the jammer can then disguise itself by transmitting symbols from Ω over a fake channel, such that it is difficult for the receiver to distinguish the authorized user's signal from the jamming interference.

For this reason, just with a power level close or equal to the signal power, disguised jamming can lead to complete transmission failure.

1.2 Jamming Resistance of General Communication Systems

For a long time, research on communication system design has been focused on capacity improvement under non-intentional interference, such as intersymbol interference, multiuser interference, and noise. Most of the communication systems today, such as GSM and OFDM, do not really have anti-jamming features. Their jamming resistance mainly relies on the diversity introduced by error control coding. On the other hand, jamming has been widely modeled as Gaussian noise. Based on the noise jamming model and the Shannon capacity formula, $C = B\log_2(1+SNR)$, an intuitive impression is that jamming is really harmful only when the jamming power is much higher than the signal power.

However, this result is only partially true. To show it, we consider *disguised jamming* [11, 13], where the jamming is highly correlated with the signal and has a power level close or equal to the signal power. *Disguised jamming can be much more harmful than noise jamming; it can reduce the system capacity to zero even when the jamming power equals the signal power.* Consider the following example:

$$y = s + J + n \tag{1.5}$$

where s is the authorized signal, J is the jamming, n is the noise independent of J and s, and y is the received signal. If J and s are taken randomly and independently from the same QPSK constellation Ω, then it can be proved [14] that *the capacity of the system is zero*! The reason behind it is that due to the *symmetricity* between the jamming and the authorized signal, the receiver is fully confused and cannot distinguish the authorized signal from jamming. Moreover, the result cannot be changed by applying the conventional bit-level channel coding. From this example, we can see that *the communication systems today are facing much more serious threats from hostile jamming than we had thought.*

1.3 Limitations with Existing Anti-Jamming Techniques

Existing work on anti-jamming system design or jamming mitigation is mainly based on spread spectrum techniques [15–20]. The spread-spectrum systems, including CDMA and frequency hopping (FH), were originally developed for secure communications in military applications. Both CDMA and FH systems possess anti-jamming and anti-interception features by exploiting frequency diversity over large spectrum [21].

In CDMA, each user is assigned a specific pseudorandom code (also known as the signature waveform) to spread its signal energy over a bandwidth N times larger. Due to the spread spectrum, CDMA is especially robust under narrowband jamming. CDMA signals cannot be recovered unless the user signature is known at the receiver and can be hidden within the noise floor, making it difficult to be detected. *In FH*, the transmitter hops in a pseudorandom manner among available frequencies according to a pre-specified PN sequence, and the receiver then operates in strict synchronization with the transmitter and remains tuned to the same center frequency [22]. When the hopping pattern is unknown, it is difficult for the adversaries to jam the FH signal. *While these systems work reasonably well for voice-centric communication, their security feature and information capacity are far from adequate and acceptable for today's high-speed multimedia wireless services* [23].

1.3.1 Inadequate Jamming Resistance Due to Security Weaknesses

The security of CDMA and FH relies on the randomness in the PN sequence. For CDMA, it is the long code used for scrambling after the spreading process. The spreading code of each user is obtained through the modulo 2 sum of the Walsh code and the long code and thus is varying in every symbol period. For FH, it is the PN sequence used for hopping frequency control. So how safe are these PN sequences? What would be the result if they are broken?

According to the Berlekamp-Massey algorithm [24], for a sequence generated from an n-stage linear feedback shift register, the characteristic polynomial and the entire sequence can be reconstructed if an eavesdropper can intercept a $2n$-bit sequence segment. Note that the characteristic polynomial is generally available to the public and then PN sequence can be recovered if an n-bit sequence segment is intercepted. That is, it is possible to break the PN sequence used in conventional CDMA and FH systems in real time with today's high-speed computing techniques.

Once the PN sequence is recovered or broken, the jammer can then launch follower jamming and disguised jamming. For CDMA, the jammer can transmit a different signal from the same constellation using the recovered spreading code of the authorized user. As a result, the authorized signal is completely jammed. For FH, the jammer can recover the hopping pattern and launch follower hamming, which jams every hopping band of the user. Again, effective communication is destroyed. As shown in Fig. 1.1, *for both CDMA and FH, recovery of the PN sequence leads to complete communication failure*. In summary, due to the security weaknesses of the PN sequences, existing CDMA and FH systems are fragile under hostile jamming, especially disguised jamming.

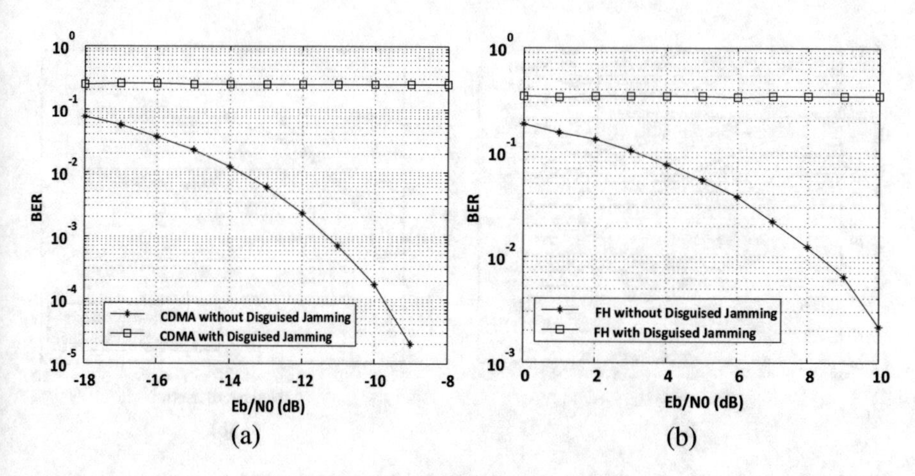

Fig. 1.1 Performance of CDMA and FH under disguised jamming

1.3.2 Low Spectral Efficiency Due to Self-Jamming and Repeated Coding

Mainly limited by multiuser interference (caused by multipath propagation and asynchronization in CDMA systems and by collision effects in FH systems) and repeated coding, the efficiency of existing spread spectrum systems is very low due to inefficient use of the total available bandwidth [23, 25, 26].

In multiple-access CDMA, all the users transmit over the same frequency band simultaneously. At the receiver, users are separated by exploiting the orthogonality between their signature waveforms. Self-jamming occurs among the users when the orthogonality is lost due to asynchronization and multipath propagation. As a result, for a CDMA system with spreading vector N, in general, only $\frac{N}{4}$ users can be supported by the system.

In multiple-access FH, each user hops independently based on its own PN sequence, and a collision occurs whenever there are two or more users transmitting over the same frequency band. In this case, the probability of bit error is generally close to 0.5 [27]. Assuming there are N_c available channels and M active users, all N_c channels are equally probable, and all users hop independently, even if each user only transmits over a single carrier, then the probability that a collision occurs is given by

$$P_h = 1 - (1 - \frac{1}{N_c})^{M-1}. \tag{1.6}$$

When N_c is large, $P_h \approx \frac{M-1}{N_c}$. This collision effect is illustrated in Fig. 1.2. Again, only a very limited number of users can be supported over the large spectrum.

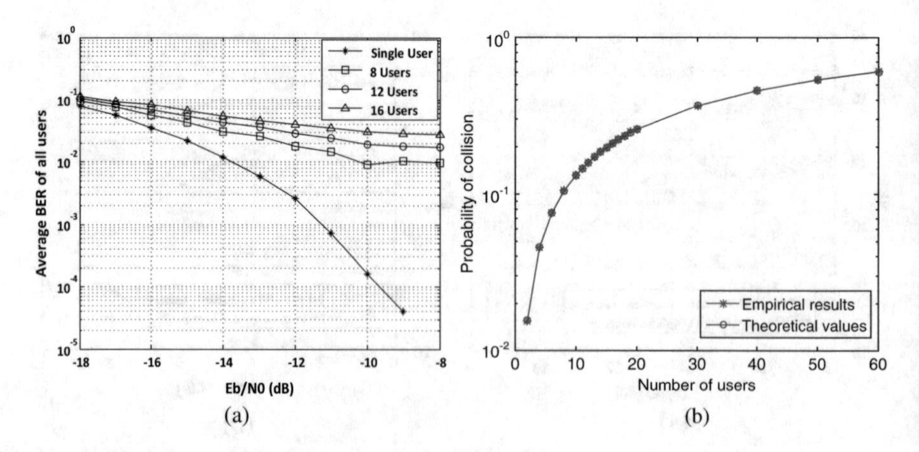

Fig. 1.2 (**a**) Self-jamming in CDMA, $N = 64$; (**b**) Collision-effect in FH, $N_c = 64$

Another observation is that in FH, for each user, the same information is transmitted at each hop, while in CDMA, the phase-shifted version of the same signal is transmitted at every chip period. Essentially, both methods use *repeated coding*, which is generally not the best coding choice from the efficiency point of view.

1.4 The Arbitrarily Varying Channel Model and Channel Capacity Under Jamming

The arbitrarily varying channel (AVC) model was first introduced by Blackwell, Breiman, and Thomasian in 1959 [28] when studying random coding. An AVC channel has unknown parameters that can change over time, and these changes may not have a uniform pattern during the signal transmission. AVC channel provides a perfect model for channels under cognitive jamming [14, 29].

The AVC channel model is generally characterized using a kernel

$$W : \mathcal{S} \times \mathcal{J} \to \mathcal{Y}, \tag{1.7}$$

where $\mathcal{S}$ is the transmitted signal space, $\mathcal{J}$ is the jamming space (i.e., the jamming is viewed as the state of the arbitrarily varying channel), and $\mathcal{Y}$ is the estimated signal space. For any $\mathbf{s} \in \mathcal{S}$, $\mathbf{j} \in \mathcal{J}$, and $\mathbf{y} \in \mathcal{Y}$, $W(\mathbf{y}|\mathbf{s}, \mathbf{j})$ denotes the conditional probability that $\mathbf{y}$ is detected at the receiver, given that $\mathbf{s}$ is the transmitted signal and $\mathbf{j}$ the jamming.

Definition 1.1 ([32]) The AVC is said to have a symmetric kernel, if $\mathcal{S} = \mathcal{J}$ and $W(\mathbf{y}|\mathbf{s}, \mathbf{j}) = W(\mathbf{y}|\mathbf{j}, \mathbf{s})$ for any $\mathbf{s}, \mathbf{j} \in \mathcal{S}$, $\mathbf{y} \in \mathcal{Y}$.

Definition 1.2 ([32]) Define $\hat{W} : \mathcal{S}\times\mathcal{S} \to \mathcal{Y}$ by $\hat{W}(\mathbf{y}|\mathbf{s}, \mathbf{s}') \triangleq \sum_{\mathbf{j}\in\mathcal{J}'} \pi(\mathbf{j}|\mathbf{s}')W(\mathbf{y}|\mathbf{s}, \mathbf{j})$, where $\pi : \mathcal{S} \to \mathcal{J}'$ is a probability matrix and $\mathcal{J}' \subseteq \mathcal{J}$. If there exists a $\pi : \mathcal{S} \to \mathcal{J}'$ such that $\hat{W}(\mathbf{y}|\mathbf{s}, \mathbf{s}') = \hat{W}(\mathbf{y}|\mathbf{s}', \mathbf{s})$, $\forall \mathbf{s}, \mathbf{s}' \in \mathcal{S}$, $\forall \mathbf{y} \in \mathcal{Y}$, then W is said to be symmetrizable.

To help elaborate the physical meaning of these concepts, symmetric and symmetrizable AVC kernels are depicted in Fig. 1.3. In an AVC with a symmetric kernel, jamming is generated from exactly the same signal space as that of the authorized signal. Even if the roles of the authorized signal and the jamming are switched, the receiver cannot detect any differences, i.e., $W(\mathbf{y}|\mathbf{s}, \mathbf{j}) = W(\mathbf{y}|\mathbf{j}, \mathbf{s})$. In an AVC with a symmetrizable kernel, jamming is generated or can be viewed as the jammer excites the main channel via an auxiliary channel $\pi : \mathcal{S} \to \mathcal{J}$, where π is essentially a probability matrix. More specifically, the input of the auxiliary channel comes from exactly the same signal space as that of the authorized signal, and it is transformed by the auxiliary channel and then imposed to the main channel. An AVC kernel is said to be symmetrizable, if there exists an auxiliary channel π, such that even if we switch the authorized signal and the input signal of the auxiliary

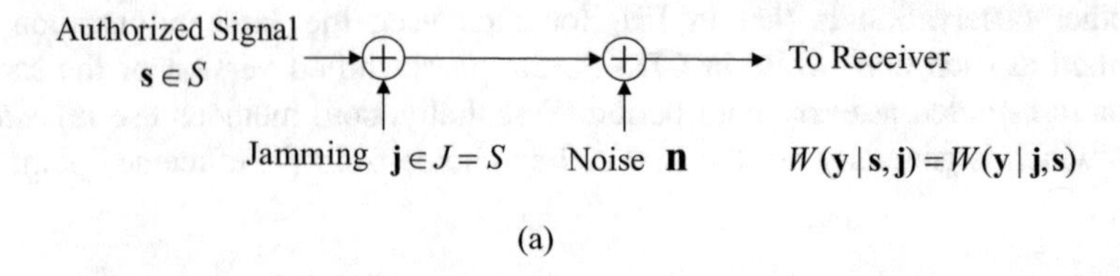

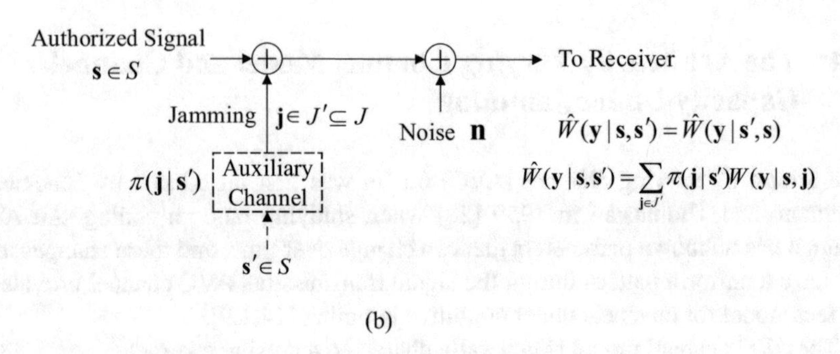

Fig. 1.3 An illustration of symmetric and symmetrizable AVC kernels. (© [2016] IEEE. Reprinted, with permission, from ref. [33]). (**a**) Symmetric kernel. (**b**) Symmetrizable kernel

channel, the receiver cannot tell any differences, i.e., $\hat{W}(\mathbf{y}|\mathbf{s}, \mathbf{s}') = \hat{W}(\mathbf{y}|\mathbf{s}', \mathbf{s})$ with $\hat{W}(\mathbf{y}|\mathbf{s}, \mathbf{s}') \triangleq \sum_{\mathbf{j} \in \mathcal{J}'} \pi(\mathbf{j}|\mathbf{s}') W(\mathbf{y}|\mathbf{s}, \mathbf{j})$. In both cases, the receiver will be confused by the disguised jamming (either generated directly or via an auxiliary channel), which is indistinguishable from the authorized signal. An interesting observation is that for an AVC kernel, being symmetric is actually a special case of being symmetrizable, where the output of the auxiliary channel equals its input.

Let I denote the mutual information between the input and the output. In the jamming-free case, the traditional Shannon channel capacity $\mathcal{C}$ is given by

$$\mathcal{C} = \max_P I(P), \tag{1.8}$$

where P denotes the statistical distribution of the input $\mathbf{s}$ and the mutual information is maximized over all possible distributions of the input. When jamming is present, the channel capacity is given by

$$\mathcal{C} = \max_P \min_\zeta I(P, \zeta), \tag{1.9}$$

where P is the statistical distribution of the input $\mathbf{s}$ and ζ the distribution of the jamming $\mathbf{j}$. Here the mutual information is maximized over all possible distributions of the input and minimized over all the possible distribution of the jammer. In other words, the user aims to maximize the channel capacity, and the jammer aims to minimize the capacity of the user.

The capacity of the AVC channel, which is characterized by (1.9), has been studied intensively in literature [29–32]. It was shown that [32] *the deterministic code capacity*[1] *of an AVC for the average probability of error is positive if and only if the AVC is neither symmetric nor symmetrizable.*

This result implies that the symmetricity between the authorized user and the malicious jammer can completely paralyze the communication system, since the receiver can no longer distinguish the user signal from the jamming interference. Obviously, for secure and effective communications under hostile jamming, an important task in system design is to break the symmetricity between the user and the jammer.

1.5 Book Overview: Anti-Jamming System Design and Capacity Analysis Under Jamming

Based on the discussion above, our goal is to design highly secure and efficient wireless communication systems under hostile jamming attacks. This is a great challenge since security is generally achieved at the cost of lower spectral efficiency. In our research on anti-jamming system design over the last decade, the approach we take is *to incorporate the time-frequency-space diversity and advanced cryptographic techniques into the physical layer transceiver design and spectrum access control, so as to achieve jamming resistance through the synthesis of secure randomness and diversity*.

The major components of this book are outlined as follows:

Chapter 2 considers *robust CDMA system design and capacity analysis under disguised jamming*, where the jammer generates a fake signal using the same spreading code, constellation, and pulse shaping filter as that of the authorized signal. *First*, we analyze the performance of conventional CDMA systems under disguised jamming and show that due to the symmetricity between the authorized signal and the jamming interference, the receiver cannot distinguish the authorized signal from jamming, leading to complete communication failure. *Second*, we explore effective CDMA system design under disguised jamming. We find that it is possible to combat disguised jamming through careful receiver design, by exploiting the time difference between the authorized signal and jamming. However, this approach only works when the time difference between the authorized signal and jamming is significant enough. A more reliable approach is to equip CDMA with secure scrambling, where we apply Advanced Encryption Standard (AES) to generate the scrambling codes. This approach performs well even if when the authorized signal and the jamming are perfectly synchronized. Based on the information theory,

[1] A deterministic code capacity is defined by the capacity that can be achieved by a communication system, when it applies only one code pattern during the information transmission. In other words, the coding scheme is deterministic and can be readily repeated by other users [32].

we show that the AVC corresponding to the conventional CDMA is symmetric, hence resulting in zero deterministic code capacity. On the other hand, the AVC corresponding to securely scrambled CDMA is not symmetrizable and hence can achieve positive deterministic code capacity under disguised jamming.

Chapter 3 considers *spectrally efficient frequency hopping (FH) system design and capacity analysis under disguised jamming*. First, we present an innovative message-driven frequency hopping (MDFH) scheme. Unlike in traditional FH where the hopping pattern of each user is determined by a preselected pseudo-random (PN) sequence, in MDFH, part of the message stream acts as the PN sequence and is transmitted through hopping frequency control. As a result, system efficiency is increased significantly since additional information transmission is achieved at no extra cost on either bandwidth or power. Quantitative performance analysis on the presented schemes demonstrates that transmission through hopping frequency control essentially introduces another dimension to the signal space, and the corresponding coding gain can increase the system efficiency by *multiple* times. Second, we analyze the performance of MDFH under hostile jamming and present an anti-jamming MDFH (AJ-MDFH) system. Note that while MDFH is robust under strong jamming, it experiences considerable performance losses under disguised jamming from sources that mimic the true signal. To overcome this limitation, in the anti-jamming MDFH (AJ-MDFH) system, we transmit a secure ID sequence along with the information stream. The ID sequence is generated through a cryptographic algorithm using the shared secret between the transmitter and the receiver, and it is then exploited by the receiver for effective signal extraction. It is shown that AJ-MDFH can effectively reduce the performance degradation caused by disguised jamming and is also robust under strong jamming. In addition, we extend AJ-MDFH to multi-carrier case, which can increase the system efficiency and jamming resistance significantly through jamming randomization and more careful spectrum usage, and can readily be used as a collision-free multiple-access system. Finally, using the arbitrarily varying channel (AVC) model, we analyze the capacity of MDFH and AJ-MDFH under disguised jamming. We show that under the worst case disguised jamming, as long as the secure ID sequence is unavailable to the jammer (which is ensured by AES), the AVC corresponding to AJ-MDFH is nonsymmetrizable. This implies that the deterministic code capacity of AJ-MDFH with respect to the average probability of error is positive. On the other hand, due to the lack of shared randomness, the AVC corresponding to MDFH is symmetric, resulting in zero deterministic capacity. We further calculate the capacity of AJ-MDFH and show that it converges as the ID constellation size goes to infinity.

Chapter 4 considers *OFDM-based highly efficient anti-jamming system design using secure dynamic spectrum access control*. First, we present a collision-free frequency hopping (CFFH) system based on the OFDMA framework and an innovative secure subcarrier assignment scheme. The secure subcarrier assignment algorithm is designed to ensure that: (i) each user hops to a new set of subcarriers in a pseudorandom manner at the beginning of each hopping period; (ii) different users always transmit on non-overlapping sets of subcarriers; and (iii) malicious users cannot predict or repeat the hopping pattern of the authorized users and hence cannot launch follower jamming attacks. That is, the CFFH scheme can effectively mitigate the jamming interference, including both random jamming and follower jamming. Moreover, CFFH has the same high spectral efficiency as that of the OFDM system and can relax the complex frequency synchronization problem suffered by conventional FH. We further enhance the anti-jamming property of CFFH by incorporating the space-time coding (STC) scheme. The enhanced system is referred to as STC-CFFH. Our analysis indicates that the combination of space-time coding and CFFH is particularly powerful in eliminating channel interference and hostile jamming interference, especially random jamming. We would like to emphasize that CFFH, although originally designed for OFDM, is not limited to OFDM and can be applied to any multiband multiaccess communication systems.

Chapter 5 considers *OFDM system design under disguised jamming*. While the collision-free frequency hopping (CFFH)-based OFDM is very effective in combating partial-band jamming, it is still fragile under disguised jamming, where the jammer mimics the characteristics of the authorized signal. For reliable transmission under disguised jamming, in this chapter, we present a securely precoded OFDM (SP-OFDM) to achieve constellation randomization. The basic idea is to introduce securely shared randomness between the legitimate transmitter and receiver at the symbol level, and hence break the symmetricity between the authorized signal and the disguised jamming. We analyze the channel capacities of both the traditional OFDM and SP-OFDM under hostile jamming using the arbitrarily varying channel (AVC) model. It is shown that the deterministic code capacity of the traditional OFDM is zero under the worst disguised jamming. On the other hand, due to the secure randomness shared between the authorized transmitter and receiver, SP-OFDM can achieve a positive capacity under disguised jamming since the AVC corresponding to SP-OFDM is not symmetrizable. A remarkable feature of the SP-OFDM scheme is that while achieving strong jamming resistance, it has the same high spectral efficiency as the traditional OFDM system. The robustness of the SP-OFDM scheme under disguised jamming is demonstrated through both theoretic and numerical analyses. In addition, we also investigate the worst jamming distribution that minimizes the channel capacity of SP-OFDM through numerical examples. It is shown that limited by the peak transmit power, the worst jamming for SP-OFDM also has a finite constellation; however, due to the inherent secure randomness in SP-OFDM, disguised jamming is no longer the worst jamming for SP-OFDM.

Chapter 6 considers *multiband transmission under jamming from a game theoretic perspective.* When a power-limited authorized user and a power-limited jammer are operating swiftly and independently over the same spectrum consisting of multiple bands, it would be impossible to track the opponent's transmission pattern. In this case, the strategic decision-making of the authorized user and the jammer can be modeled as a two-party zero-sum game, where the payoff function is the capacity that can be achieved by the authorized user in the presence of the jammer. In this chapter, first, we investigate the game under additive white Gaussian noise (AWGN) channels. We explore the possibility for the authorized user or the jammer to randomly utilize part (or all) of the available spectrum and/or apply nonuniform power allocation. It is found that under AWGN channels, either for the authorized user to maximize its capacity or for the jammer to minimize the capacity of the authorized user, the best strategy is to distribute the transmission power or jamming power uniformly over all the available spectra. The minimax capacity can be calculated based on the channel bandwidth and the signal-to-jamming and noise ratio, and it matches with the Shannon channel capacity formula. Second, we consider frequency-selective fading channels. We characterize the dynamic relationship between the optimal signal power allocation and the optimal jamming power allocation in the minimax game and present an efficient two-step water-pouring algorithm to find the optimal power allocation schemes for both the authorized user and the jammer.

Chapter 7 *summarizes what we have learned from our study on jamming and anti-jamming and looks into future research directions.* We summarize the jamming models, the essential techniques in anti-jamming system design, the innovative anti-jamming systems, and the ultimate game between smart transmission and cognitive jamming. We also share our thoughts on some future research directions on jamming and anti-jamming in 5G IoT systems and 5G wireless network design and performance evaluation under hostile environments.

References

1. R. Pickholtz, D. Schilling, and L.B.Milstein, "Theory of spread spectrum communications – a tutorial," *IEEE Transactions on Communications*, vol. 30, pp. 855–884, May 1982.
2. M. Pursley and W. Stark, "Performance of reed-solomon coded frequency-hop spread-spectrum communications in partial-band interference," *IEEE Transactions on Communications*, vol. 33, pp. 767–774, Aug 1985.
3. B. Levitt, "FH/MFSK performance in multitone jamming," vol. 3, no. 5, pp. 627–643, 1985.
4. A. Alahmadi, Z. Fang, T. Song, and T. Li, "Subband puea detection and mitigation in ofdm-based cognitive radio networks," *IEEE Transactions on Information Forensics and Security*, vol. 10, no. 10, pp. 2131–2142, Oct 2015.
5. Y. R. Chien, "Design of gps anti-jamming systems using adaptive notch filters," *IEEE Systems Journal*, vol. 9, no. 2, pp. 451–460, June 2015.
6. J.-W. Moon, J. Shea, and T. Wong, "Jamming estimation on block-fading channels," in *Proc. IEEE Military Communications Conference*, vol. 3, Oct. 31- Nov. 3 2004, pp. 1310–1316.
7. N. Pronios and A. Polydoros, "Jamming optimization in fully-connected, spread-spectrum networks," in *Proc. IEEE Military Commun. Conf.* IEEE, pp. 65–70.
8. M. Pursley and J. Skinner, "Turbo product coding in frequency-hop wireless communications with partial-band interference," in *Proc. IEEE Military Communications Conference*, vol. 2, Oct 2002, pp. 774–779.
9. L. Zhang, J. Ren, and T. Li, "Time-varying jamming modeling and classification," *IEEE Transactions on Signal Processing*, vol. 60, no. 7, pp. 3902–3907, July 2012.
10. L. Zhang, H. Wang, and T. Li, "Jamming resistance reinforcement of message-driven frequency hopping," in *Proc. Intl. Conf. Acoust., Speech, Signal Processing*, Mar. 2010, pp. 3974–3977.
11. ——, "Anti-Jamming Message-Driven Frequency Hopping: Part I — System Design," *IEEE Transactions on Wireless Communications*, pp. 70–79, 2013.
12. L. Zhang and T. Li, "Anti-Jamming Message-Driven Frequency Hopping: Part II — Capacity Analysis Under Disguised Jamming," *IEEE Transactions on Wireless Communications*, pp. 80–88, 2013.

13. M. Medard, "Capacity of correlated jamming channels," in *Allerton Annual Conf. Commun., Control Comput.*, 1997.
14. A. Lapidoth and P. Narayan, "Reliable communication under channel uncertainty," *IEEE Transactions on Information Theory*, vol. 44, no. 6, pp. 2148–2177, Oct 1998.
15. R. Alred, "Naval radar anti-jamming technique," *Journal of the Institution of Electrical Engineers – Part III A: Radiolocation*, vol. 93, no. 10, pp. 1593–1601, 1946.
16. M. Amin, "Interference mitigation in spread spectrum communication systems using time-frequency distributions," *IEEE Transactions on Signal Processing*, vol. 45, no. 1, pp. 90–101, 1997.
17. W. Sun and M. Amin, "A self-coherence anti-jamming gps receiver," *IEEE Transactions on Signal Processing*, vol. 53, no. 10, pp. 3910–3915, 2005.
18. I. Bergel, E. Fishler, and H. Messer, "Narrowband interference mitigation in impulse radio," *IEEE Transactions on Communications*, vol. 53, no. 8, pp. 1278–1282, 2005.
19. Y. Wu, B. Wang, K. Liu, and T. Clancy, "Anti-jamming games in multi-channel cognitive radio networks," *IEEE Journal on Selected Areas in Communications*, vol. 30, no. 1, pp. 4–15, 2012.
20. C. Popper, M. Strasser, and S. Capkun, "Anti-jamming broadcast communication using uncoordinated spread spectrum techniques," *IEEE Journal on Selected Areas in Communications*, vol. 28, no. 5, pp. 703–715, 2010.
21. R. Dixon, *Spread Spectrum Systems with Commercial Applications*, 3rd ed.John Wiley & Son, Inc, 1994.
22. F. Dominique and J. Reed, "Robust frequency hop synchronisation algorithm," *Electronics Letters*, vol. 32, pp. 1450–1451, Aug. 1996.
23. T. Li, J. Ren, Q. Ling, and W. Liang, "Physical Layer Built-in Security Analysis and Enhancement of CDMA Systems," in *Proc. Conference on Information Sciences and Systems*, University of Princeton, Princeton, NJ, Mar. 2004.
24. J. Massey, "Shift-Register Synthesis and BCH Decoding," *IEEE Transactions on Information Theory*, vol. 15, pp. 122–127, Jan. 1969.
25. Q. Ling, T. Li, and Z. Ding, "A Novel Concept: Message Driven Frequency Hopping (MDFH)," in *Proc. IEEE International Conference on Communications*, 2007.
26. M. Zhang, C. Carroll, and A. H. Chan, "Analysis of IS-95 CDMA Voice Privacy," in *Selected Areas in Cryptography*, 2000, pp. 1–13.
27. T. Rappaport, *Wireless Communications–Principles and Practices*, 2nd ed.Prentice Hall, 2002.
28. D. Blackwell, L. Breiman, and A. J. Thomasian, "The capacities of certain channel classes under random coding," *Ann. Math. Statist.*, vol. 31, no. 3, pp. 558–567, 09 1960. [Online]. Available: https://doi.org/10.1214/aoms/1177705783
29. I. Csiszar and P. Narayan, "The capacity of the arbitrarily varying channel revisited: positivity, constraints," *IEEE Transactions on Information Theory*, vol. 34, no. 2, pp. 181–193, Mar 1988.
30. R. Ahlswede and V. Blinovsky, "Classical capacity of classical-quantum arbitrarily varying channels," *IEEE Transactions on Information Theory*, vol. 53, no. 2, pp. 526–533, Feb 2007.
31. V. Blinovsky, P. Narayan, and M. Pinsker, "Capacity of the arbitrarily varying channel under list decoding," *Problems of Information Transmission*, vol. 31, no. 2, pp. 99–113, 1995.
32. T. Ericson, "Exponential error bounds for random codes in the arbitrarily varying channel," *IEEE Transactions on Information Theory*, vol. 31, no. 1, pp. 42–48, Jan 1985.
33. T. Song, W. E. Stark, T. Li, and J. K. Tugnait "Optimal multiband transmission under hostile jamming," *IEEE Transactions on Communications*, vol. 64, no. 9, pp. 4013–4027, Sept 2016.

Chapter 2
Enhanced CDMA System with Secure Scrambling

2.1 Introduction

2.1.1 CDMA and Its Security

Existing work on anti-jamming system design or jamming mitigation is mainly based on spread-spectrum techniques [1, 2]. The spread-spectrum systems, including code division multiple access (CDMA) and frequency hopping (FH), were originally developed for secure communications in military applications. Both CDMA and FH systems possess anti-jamming and anti-interception features by exploiting frequency diversity over large spectrum.

In CDMA, each user is assigned a specific pseudo-random code (also known as the signature waveform) to spread its signal over a bandwidth N times larger. Due to the processing gain resulted from the spread-spectrum technique, CDMA is especially robust under narrowband jamming and works well under low SNR levels [3]. Hidden within the noise floor, CDMA signals are difficult to be detected and cannot be recovered unless the user signature is known at the receiver. For these reasons, CDMA has been widely used in both civilian and military applications, such as 3GPP UMTS [4] and GPS [5].

The security of CDMA largely relies on the randomness in the PN sequence. For CDMA, the spreading code of each user is obtained through the modulo-2 sum of the Walsh code and the long code and thus is varying in every symbol period. However, according to the Berlekamp-Massey algorithm [6], for a sequence generated from an n-stage linear feedback shift register, the characteristic polynomial and the entire sequence can be reconstructed if an eavesdropper can intercept a $2n$-bit sequence segment. Note that the characteristic polynomial is generally available to the public and then PN sequence can be recovered if an n-bit sequence segment is intercepted. That is, it is possible to break the PN sequence used in the conventional CDMA systems in real time with today's high-speed computing techniques [7]. Once the PN sequence is recovered or broken, the jammer can generate a fake signal

T. Li et al., *Wireless Communications Under Hostile Jamming: Security and Efficiency*, https://doi.org/10.1007/978-981-13-0821-5_2

using the same spreading code, constellation, and pulse shaping filter as that of the authorized signal. This is known as the *disguised jamming* [8–10] for CDMA.

In this chapter, we will analyze the performance of conventional CDMA systems under disguised jamming and then introduce two innovative approaches to combat disguised jamming through robust receiver design and secure scrambling.

2.1.2 Existing Work

Regarding the jamming issues of CDMA systems, we are particularly interested in three critical questions: (1) What is the capacity of CDMA under jamming in general? (2) What is the most effective jamming for CDMA? (3) How secure are the existing CDMA systems? In this subsection, we would like to review several most relevant research papers in literature, from which we can find some good answers.

2.1.2.1 Capacity of CDMA Systems Under Jamming

Nikjah et al. [11] investigated the capacity of general CDMA systems under jamming. They considered a memoryless CDMA jamming channel described for each channel degree of freedom[1] by the baseband model:

$$Y = gX + \Theta J + Z, \tag{2.1}$$

where Y is the soft output of the channel. Random variable (RV) g is the channel complex gain. It is assumed that $g = 1$ in static channels and $|g| \sim \text{Nakagami}(m, 1)$ in fading channels, where $\text{Nakagami}(m, \Omega)$ denotes the Nakagami-m distribution.[2] The complex variable X in (2.1) is the transmitted symbol with mean zero and variance E. The discrete RV Θ is the jammer state with the following probability function:

$$p_\Theta(\theta) = \begin{cases} 1-\rho, & \theta = 0, \\ \rho, & \theta = 1, \end{cases} \tag{2.2}$$

where $\rho \in (0, 1]$. When the channel is being jammed, $\Theta = 1$; otherwise, $\Theta = 0$. In other words, it is assumed that the channel is jammed in a fraction ρ of its available degrees of freedom. The RV J in (2.1) is the jamming interference which is assumed to be a zero-mean, circularly symmetric complex Gaussian (CSCG) RV with variance N_J/ρ, where N_J is the jamming power per degree of freedom if

[1] Channel degrees of freedom are independent time, frequency, or space dimensions of the channel used for data transmission.

[2] https://en.wikipedia.org/wiki/Nakagami_distribution

the jammer is omnipresent, i.e., if it contaminates all available degrees of freedom. Lastly, variable Z in (2.1) is a zero-mean CSCG RV with variance σ_Z^2 accounting for ambient noise and multiaccess interference (MAI).

In [11], the capacity of CDMA was calculated under different jamming scenarios, which are categorized by: (a) cooperative or noncooperative[3] CDMA, (b) whether the jammer state information (JSI) is known to the receiver or not, and (c) static or fading channels. For noncooperative CDMA systems with static channels and JSI known to the receiver, it was shown that the capacity can be calculated as:

$$C = (1-\rho)\log\left(1+\frac{E}{\sigma_Z^2}\right)+\rho\log\left(1+\frac{E}{N_J/\rho+\sigma_Z^2}\right). \tag{2.3}$$

If interested, the reader is referred to [11] for the capacity calculation for all the other cases.

Based on the capacity analysis, the authors of [11] came up with an important conclusion: *in order to minimize the average capacity of the authorized user, the jammer should spread its energy evenly over all degrees of freedom.*

2.1.2.2 Deterministic/Random Codes and Most Effective Jamming

In [12], Ericson discussed the arbitrarily varying channel (AVC) model and the capacity of a general communication system under AVC with deterministic codes and random codes, respectively. We have introduced the AVC model in Sect. 1.4, so here we will briefly revisit deterministic code, random code, and how they are connected to the channel capacity of a communication system under jamming.

A deterministic code stands for a code pattern which is constant during information transmission. Whereas, random codes change constantly, and nobody except the authorized parties could know or predict what the code will be at a future time. In [12], it was described as: "A random code can be thought of as a code chosen at random from a given ensemble of deterministic codes. It is an essential assumption that the jammer is unaware of which code has actually been chosen (although he might be aware of the ensemble, as well as of the random mechanism used for the selection)."

Under the AVC channel model, where a jammer is assumed to be present, "the performance of random codes, in general, exceeds that of deterministic codes." More specifically, if the AVC kernel is symmetric or symmetrizable (see Sect. 1.4 for detailed definitions), the deterministic code capacity of a communication system under AVC is zero. The underlying reason is that, with a symmetric or symmetrizable AVC kernel, the signal space and jamming space are identical (without or with proper mapping of spaces), such that the receiver will not be able to tell the

[3] A CDMA channel is called noncooperative when the user transmissions are not coordinated with each other and the receivers treat the signals from the other users as noise.

difference when the signal and jamming are mixed (added) together in the channel. In other words, "symmetric or symmetrizable" implies that jamming and authorized signal are not distinguishable at the receiver. In this case, the error probability will always be lower bounded. More specifically, the symbol error probability is lower bounded by $\frac{M-1}{2M}$, where M is the size of the constellation.

However, the situation above can be changed by random codes, which essentially break the symmetricity between the authorized user and the jammer. In other words, the channel capacity would no longer be zero, when random codes are applied. Note that random codes are not only random but also well protected from interception by any malicious users. Hence, whether a communication system under jamming can have a positive capacity would quite depend on what codes it adopts, deterministic or random.

Returning to our original CDMA problem, for a jammer who wants to limit the communication of the authorized user (potentially reducing the channel capacity to zero), the most effective way of jamming is to find out the code pattern and mimic the transmission behavior of the authorized user. This kind of jamming (also known as disguised jamming) is a disaster to a CDMA system, as it will lead to a zero capacity for the authorized user. Can a jammer find out the code pattern of existing CDMA systems and launch disguised jamming? This largely depends on the built-in security of existing CDMA systems. As we will show in the following subsection, the built-in security of existing CDMA systems is far from adequate.

2.1.2.3 Built-In Security Analysis of Existing CDMA Systems

In our previous work [7], we have analyzed the security of two existing CDMA systems: IS-95 and 3GPP UMTS. Since the channelization codes (typically Walsh codes) adopted in the spreading process are typically short and easy to generate, the physical layer built-in security of CDMA systems mainly relies on the long pseudo-random scrambling code, also known as long code. However, it has been found that the scrambling codes of existing CDMA systems are not sufficiently secure.

In IS-95, the long-code generator consists of a 42-bit number called long-code mask and a 42-bit linear feedback shift register (LFSR) specified by a polynomial of degree 42. Each chip of the long code is generated by the modulo-2 inner product of a 42-bit mask and the 42-bit state vector of the LFSR. Let $M = [m_1, m_2, \ldots, m_{42}]$ denote the 42-bit mask and $S(t) = [s_1(t), s_2(t), \ldots, s_{42}(t)]$ denote the state of the LFSR at time instant t. The long-code sequence $c(t)$ at time t can thus be represented as:

$$c(t) = m_1 s_1(t) + m_2 s_2(t) + \ldots + m_{42} s_{42}(t), \tag{2.4}$$

where the additions are modulo-2 additions.

As it is well known, for a sequence generated from an n-stage linear feedback shift register, if an eavesdropper can intercept a $2n$-bit sequence segment, then the characteristic polynomial and the entire sequence can be reconstructed according to the Berlekamp-Massey algorithm [6]. This leaves an impression that the maximum complexity to recover the long-code sequence $c(t)$ is $O(2^{84})$. However, for IS-95, since the characteristic polynomial is known to the public, an eavesdropper only needs to obtain 42 bits of the long-code sequence to determine the entire sequence. That is, the maximum complexity to recover the long-code sequence $c(t)$ of IS-95 is only $O(2^{42})$.

In the 3GPP UMTS standard, Gold codes generated from two generator polynomials of degree 18 are used as scrambling codes. It is shown that the maximum complexity to recover the scrambling code of a 3GPP UMTS system based on ciphertext-only attack is $O(2^{36})$. This implies that the physical layer built-in security of the 3GPP UMTS is actually weaker than that of the IS-95 system.

Once the long-code sequence is recovered, the desired user's signal can be recovered through signal separation and extraction techniques. If the training sequence is known, simple receivers, for example, the Rake receiver, can be used to extract the desired user's signal. Even if the training sequence is unknown, the desired user's signal may still be recovered through blind multiuser detection and signal separation algorithms [13–15].

More importantly, with the long-code sequence recovered, a jammer can easily launch disguised jamming by mimicking the authorized user's transmission behavior, which will lead to zero user capacity according to what we have discussed in Sect. 2.1.2.2.

2.2 System Model and Problem Identification

2.2.1 System Model

We consider an individual user in a typical CDMA system. Assuming the processing gain is N, namely, there are N chips per symbol. Let

$$\mathbf{c} = [c_0, c_1, \ldots, c_{N-1}] \tag{2.5}$$

denote the spreading code, in which $c_n = \pm 1, \ \forall n$. In the isolated pulse case, the general baseband signal of the spreading sequence can be represented as:

$$c(t) = \sum_{n=0}^{N-1} c_n g(t - nT_c), \tag{2.6}$$

where $g(t)$ is the pulse shaping filter and T_c the chip period, and we assume:

$$\frac{1}{T} \int_0^T c^2(t)dt = 1, \tag{2.7}$$

where $T = NT_c$ is the symbol period.

Let Ω be the constellation and $u_k \in \Omega$ the kth symbol to be transmitted. The spread chip-rate signal can be expressed as:

$$q_n = u_k c_{n-kN}, \tag{2.8}$$

where $k = \lfloor \frac{n}{N} \rfloor$. The successive scrambling process is achieved by

$$z_n = q_n e_n = u_k c_{n-kN} e_n, \tag{2.9}$$

where $e_n = \pm 1$ is a pseudo-random chip-rate scrambling sequence. After pulse shaping, the transmitted signal would then be:

$$s(t) = \sum_{n=-\infty}^{\infty} u_k c_{n-kN} e_n g(t - nT_c). \tag{2.10}$$

Note that c_n, e_n, and $g(t)$ are real-valued, while u_k can be complex depending on the constellation Ω.

For an AWGN channel, the received signal can be written as:

$$y(t) = s(t) + n(t) = \sum_{n=-\infty}^{\infty} u_k c_{n-kN} e_n g(t - nT_c) + n(t), \tag{2.11}$$

where $n(t)$ is the white Gaussian noise.

To recover the transmitted symbols, the CDMA receiver first descrambles the received signal by multiplying a locally generated and synchronized copy of the scrambling sequence, e_n. Afterward, the received signal will be reduced to:

$$y(t) = \sum_{n=-\infty}^{\infty} u_k c_{n-kN} g(t - nT_c) + n(t). \tag{2.12}$$

Without loss of generality, we consider the recovery of the symbol indexed by $k = 0$ and omit the subscript k in u_k. The corresponding signal of interest would be constrained within $t = [0, T)$, and following the definition in (2.6), we have:

$$r(t) = \sum_{n=0}^{N-1} u c_n g(t - nT_c) + n(t) = uc(t) + n(t). \tag{2.13}$$

Performing the despreading process, and following (2.7), the CDMA receiver estimates the transmitted symbol as:

$$\hat{u} = \frac{1}{T}\int_0^T r(t)c(t)dt = u + \frac{1}{T}\int_0^T n(t)c(t)dt. \tag{2.14}$$

We can observe from the process above that it is impossible to recover the transmitted symbols without knowing the user's spreading code and scrambling code. This is known as a *built-in security* feature of the CDMA systems. In the following subsection, we will discuss the security level of several typical CDMA systems and show that disguised jamming, which mimics the authorized signal, can severely jeopardize the CDMA systems and, in the worst case, leads to complete communication failure.

2.2.2 Problem Identification

Since the spreading codes are generally short and easy to generate, the physical layer built-in security of typical CDMA systems mainly relies on the long pseudo-random scrambling sequence, also known as long code, e.g., in IS-95, 3GPP as well as the military GPS. However, it was shown in [7] that the long scrambling codes used by IS-95 or 3GPP UMTS can be cracked with reasonably high computational complexity. In fact, the maximum complexity to recover the long scrambling codes in IS-95 and 3GPP UMTS is only $O(2^{42})$ and $O(2^{36})$ [7], respectively. As another example, the civilian GPS even makes its codes public to attract potential users for global competitiveness.

The weakly secured or even public spreading/scrambling codes leave considerable room for malicious users to launch disguised jamming [8–10] toward the authorized signal. The jammer can mimic the authorized signal by generating fake symbols over the cracked or already known codes. With complete knowledge of the code information and the pulse shaping filter, the jammer can launch disguised jamming, which has the similar characteristics as the authorized signal, except that the fake symbol can only be randomly chosen out of Ω. Moreover, there may be small timing and amplitude differences between the authorized signal and the disguised jamming due to non-ideal estimation at the jammer side.

Let $v \in \Omega$ denote the fake symbol, τ the small timing difference, and γ the amplitude ratio of the disguised jamming to the authorized signal. Then, the disguised jamming can be modeled as:

$$j(t) = v\gamma c(t - \tau). \tag{2.15}$$

Taking both the noise and disguised jamming into account, and following (2.13), the received signal can be written as:

$$r(t) = s(t) + j(t) + n(t) = uc(t) + v\gamma c(t - \tau) + n(t), \tag{2.16}$$

where $n(t)$ is the noise.

An important observation is that *the conventional CDMA receiver as indicated in* (2.14) *would fail under disguised jamming.* In fact, replacing the received signal $r(t)$ in (2.14) with (2.16), and following (2.7), we have:

$$\hat{u} = u + v\gamma \frac{1}{T}\int_0^T c(t-\tau)c(t)dt + \frac{1}{T}\int_0^T n(t)c(t)dt. \tag{2.17}$$

As can be seen, the symbol estimation would be considerably influenced by the second term in the RHS of (2.17), which is introduced by disguised jamming, especially when τ is small (e.g., $|\tau| < T_c$) and $\gamma \approx 1$. In the worst case, when $\tau = 0$ and $\gamma = 1$, (2.17) is reduced to a very simple form:

$$\hat{u} = u + v + \frac{1}{T}\int_0^T n(t)c(t)dt. \tag{2.18}$$

We can now apply the error probability analysis result in [8], in which it was shown that in this case, the probability of symbol error, $\mathcal{P}_s$, would be lower bounded by:

$$\mathcal{P}_s \geq \frac{M-1}{2M}, \tag{2.19}$$

where M is the constellation size of Ω. An intuitive explanation is that if the authorized symbol "u" and the fake one "v" are distinct, the receiver would have to guess between them as indicated in (2.18). Note that (i) the error probability of a random guess between two symbols is $\frac{1}{2}$ and (ii) the two symbols randomly and independently selected out of Ω by the authorized transmitter and disguised jammer differ with a probability of $\frac{M-1}{M}$. Combining (i) and (ii), it then follows that $\mathcal{P}_s \geq \frac{M-1}{2M}$.

The lower bound in (2.19) sets up a limit for the error probability performance of CDMA systems under the worst-case disguised jamming (i.e., $\tau = 0$ and $\gamma = 1$), which implies that the CDMA communication is completely paralyzed. We intend to solve this problem in two separate cases: (i) For CDMA systems with public codes which cannot be concealed for some reason (e.g., in civilian GPS), in Sect. 2.3, we mitigate the disguised jamming through robust receiver design. The underlying idea is that the worst disguised jamming can hardly be launched, since the disguised jammer cannot really capture the exact timing and amplitude information of the authorized signal. In a more practical case with regular disguised jamming, we found that it is possible to mitigate the jamming considerably by taking the timing difference τ and amplitude ratio γ into account in the receiver design. (ii) For CDMA systems which allow code concealment, in Sect. 2.4, we combat disguised jamming using secure scrambling, which essentially enhances the security of the scrambling codes and hence breaks the symmetricity between the authorized user and the jammer.

2.3 Jamming Mitigation with Robust Receiver Design

For CDMA systems in which the codes cannot be concealed (e.g., civilian GPS) or the transmitters can hardly be upgraded (e.g., satellites in sky), we present an efficient way to mitigate disguised jamming by robust receiver design. With the public or easily accessed codes, disguised jamming can hardly be prevented. However, the disguised jammer cannot really capture the exact timing and amplitude information of the authorized signal, so small timing and amplitude differences between the authorized signal and disguised jamming may exist. As a result, it is possible to recover the transmitted symbols aided by proper jamming estimation. In this section, we estimate the jamming parameters as well as the authorized symbol using the minimum mean square error (MMSE) criterion. Unlike traditional MSE between the received signal and transmitted signal, the MSE here is calculated between the received signal and jammed signal, which is the sum of the authorized signal and the disguised jamming.

Following (2.10) and (2.11), the aforementioned MSE can be calculated as:

$$\begin{aligned} & J(u, v, \tau, \gamma) \\ =& \frac{1}{T}\int_0^T |r(t) - s(t) - j(t)|^2 dt \\ =& \frac{1}{T}\int_0^T |r(t) - uc(t) - v\gamma c(t-\tau)|^2 dt \\ =& \frac{1}{T}\int_0^T |r(t) - uc(t)|^2 dt - \frac{\gamma v^*}{T}\int_0^T [r(t) - uc(t)]c(t-\tau)dt \\ & - \frac{\gamma v}{T}\int_0^T [r(t) - uc(t)]^* c(t-\tau)dt + \frac{\gamma^2 |v|^2}{T}\int_0^T c^2(t-\tau)dt, \end{aligned} \tag{2.20}$$

where $(\cdot)^*$ denotes the complex conjugate. Since $c(t)$ is T-periodic, following (2.7), we have $\frac{1}{T}\int_0^T c^2(t-\tau)dt = \frac{1}{T}\int_0^T c^2(t)dt = 1$. If we further denote

$$A(u, \tau) = \frac{1}{T}\int_0^T [r(t) - uc(t)]c(t-\tau)dt, \tag{2.21}$$

the MSE can be rewritten as:

$$J(u, v, \tau, \gamma) = \frac{1}{T}\int_0^T |r(t) - uc(t)|^2 dt - \gamma v^* A(u, \tau) - \gamma v A^*(u, \tau) + \gamma^2 |v|^2. \tag{2.22}$$

Thus, the problem can be formulated as minimizing (2.22) by finding the optimal u, v, τ, and γ, i.e.:

$$\{\hat{u}, \hat{v}, \hat{\tau}, \hat{\gamma}\} = \arg\min_{u,v,\tau,\gamma} J(u, v, \tau, \gamma). \tag{2.23}$$

To minimize (2.22), one necessary condition is that its partial derivatives regarding v and γ are zero. Note that when z is a complex variable, we have $\frac{\partial z}{\partial z} = 0$, $\frac{\partial z^*}{\partial z} = 2$ and $\frac{\partial |z|^2}{\partial z} = 2z$. Hence:

$$\begin{cases} \dfrac{\partial J}{\partial v} = -2\gamma A(u, \tau) + 2\gamma^2 v = 0, & (2.24a) \\ \dfrac{\partial J}{\partial \gamma} = -v^* A(u, \tau) - v A^*(u, \tau) + 2\gamma |v|^2 = 0, & (2.24b) \end{cases}$$

from which we can get:

$$\gamma = \frac{A(u, \tau)}{v} = \frac{A^*(u, \tau)}{v^*}. \tag{2.25}$$

Substituting (2.25) into (2.22), the MSE can be reduced to:

$$J = \frac{1}{T} \int_0^T |r(t) - uc(t)|^2 dt - |A(u, \tau)|^2, \tag{2.26}$$

which is a function depending only on u and τ.

In numerical solution search, limited by the time resolution, τ becomes discrete and thus has only finite possible values with $|\tau| < T_c$. In this way, an exhaustive search[4] on τ and u would be feasible and also an effective approach to minimize (2.26). Let $\hat{u}$ and $\hat{\tau}$ be the solution pair that minimizes (2.26); following (2.25), the amplitude ratio can be estimated as:

$$\hat{\gamma} = \frac{|A(\hat{u}, \hat{\tau})|}{|v|}. \tag{2.27}$$

For a constant-modulus constellation (e.g., PSK), $|v|$ is readily available since it holds constant for all $v \in \Omega$. For nonconstant-modulus constellation, the amplitude ratio cannot be exactly drawn. This is because that from (2.25), we can only determine $\hat{v}\hat{\gamma} = A(\hat{u}, \hat{\tau})$, which cannot yield a specific $\hat{\gamma}$ when the amplitude of the jamming symbol is not specifically available. However, in this case, we can obtain a range for $\hat{\gamma}$. More specifically, if $B_1 \leq |v| \leq B_2$ for $v \in \Omega$, then we have:

[4]Generally, it would be sufficient to perform an exhaustive search for regular time resolution with a practical sampling rate; however, for high time resolution, we suggest the usage of state-of-the-art iterative optimization methods, e.g., Newton's method.

$$\frac{|A(\hat{u}, \hat{\tau})|}{B_2} \leq \hat{\gamma} \leq \frac{|A(\hat{u}, \hat{\tau})|}{B_1}. \tag{2.28}$$

Discussions

(1) The major differences between the estimation of disguised jamming and that of multipath signals [16, 17] lie in: (i) Multipath signals always contain the same symbol as the primary signal (which is the signal going through the line-of-sight path), while the symbol carried by disguised jamming is chosen independently from the authorized signal; (ii) multipath signals are generally much weaker than the primary signal, while disguised jamming maintains a similar power level as the authorized signal; and (iii) multipath signals always arrive at the receiver after the primary signal, while disguised jamming can have either a leading or lagging phase compared with the authorized signal.

(2) Although we primarily focus on recovering the authorized symbols under disguised jamming here, however, the information obtained from the MMSE receiver can be used for jamming detection and evaluation. The estimated amplitude ratio can be used as a metric to determine whether a disguised jammer is present or not by comparing it with an appropriate threshold. Through cooperation of multiple receivers, it is also possible to locate the disguised jammer by exploiting the estimated timing differences between the jamming interference and the authorized signal.

2.4 Jamming Mitigation with Secure Scrambling

As can be seen in Sect. 2.2, the physical layer security of most CDMA systems largely relies on the scrambling process. For CDMA systems whose scrambling codes are not adequately secured, to prevent the disguised jamming, we generate the scrambling sequence using the Advanced Encryption Standard (AES), also known as Rijndael.

2.4.1 AES-Based Secure Scrambling

Rijndael was identified as the new AES in October 2, 2000. Rijndael's combination of security, performance, efficiency, ease of implementation, and flexibility makes it an appropriate selection for the AES. Rijndael is a good performer in both hardware and software across a wide range of computing environments. Its low memory requirements make it very well suited for restricted-space environments such as mobile handset to achieve excellent performance. More details on AES can be found in [18].

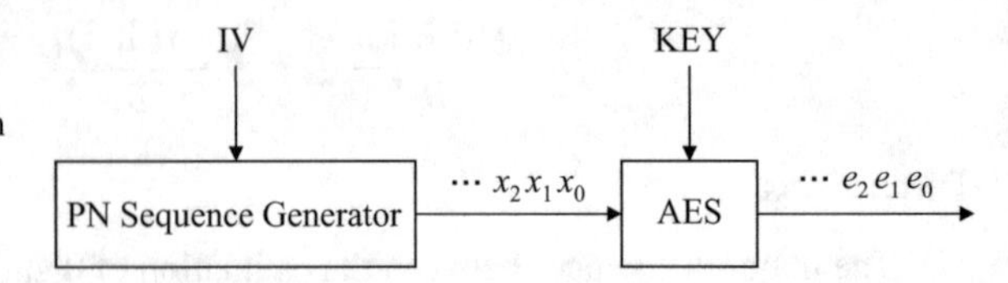

Fig. 2.1 Secure scrambling sequence generation. (© [2016] IEEE. Reprinted, with permission, from ref. [19])

The secure scrambling scheme aims to increase the physical layer built-in security of CDMA systems and prevent exhaustive key search attack while minimizing the changes required to the existing standards. As shown in Fig. 2.1, the secure scrambling sequence is generated through two steps: first, generate a pseudo-noise (PN) sequence, and then encrypt the sequence with the AES algorithm. More specifically, a PN sequence is first generated using a PN sequence generator with a secure initialization vector (IV), where the PN sequence generator is typically a linear feedback shift register (LFSR) or Gold sequence generator; subsequently the PN sequence is encrypted by the AES algorithm block by block secured by a secret encryption key, which is shared between the legitimate communication parties.

The secure scrambling process can be summarized as follows:

1. The communication parties share a common initial vector (IV) for the PN sequence generator and an L-bit ($L = 128$, 192, or 256) common secret encryption key;
2. The long scrambling sequence is generated through encryption of a particular segment of the sequence generated from the PN sequence generator using the shared secret key;
3. The scrambling process is realized by adding the scrambling sequence to the chip-rate spread signal.

2.4.2 Security Analysis

To eavesdrop on the transmitted data or launch disguised jamming, the malicious user has to intercept the secure scrambling sequence used by the legitimate users. Hence, the security of the scrambling process lies in how difficult it is to crack the encrypted scrambling sequence. In this subsection, we use data encryption standard (DES) [20] as a benchmark to evaluate the security of the secure scrambling, which is essentially ensured by AES. We compare the number of possible keys of AES, of DES, and that of the typical CDMA scrambling sequences. The number of keys determines the effort required to crack the cryptosystem by trying all possible keys.

The most important reason for DES to be replaced by AES is that it is becoming possible to crack DES through exhaustive key search. Single DES uses 56-bit encryption key, which means that there are approximately 7.2×10^{16} possible DES keys. In the late 1990s, specialized "DES cracker" machines were built, and they could recover a DES key after a few hours. In other words, by trying all possible keys, the hardware could determine which key was used to encrypt a message. Compared with DES, IS-95 has only 42-bit shared secret (approximately 4.4×10^{12} possible keys), and 3GPP UMTS has even lower security with 36-bit shared secret (approximately 6.9×10^{10} possible keys). This makes it possible to break these low-security scrambling sequences almost in real time through exhaustive key search.

On the other hand, AES specifies three key sizes: 128, 192, and 256 bits. In decimal terms, this means that approximately there are:

1. 3.4×10^{38} possible 128-bit keys;
2. 6.2×10^{57} possible 192-bit keys;
3. 1.1×10^{77} possible 256-bit keys.

Thus, if we choose $L = 128$, then there are on the order of 10^{21} times more AES 128-bit keys than DES 56-bit keys. Assuming that one could build a machine that could recover a DES key in a second (i.e., try 2^{56} keys per second), as we can see, this is a very ambitious assumption and far from what we can do today, and then it would take that machine approximately 149 thousand billion (149 trillion) years to crack a 128-bit AES key.

Security measurement through the number of all possible keys is based on the assumption that the attacker has no easy access to the secret encryption key; therefore, the attacker has to perform an exhaustive key search in order to break the system. As is well known, the security of AES is based on the infeasible complexity in recovering the encryption key. Currently, no weakness has been detected for AES; thus, exhaustive key search is still being recognized as the most effective method in recovering the encryption key and breaking the cryptosystem. Based on the calculation above, as long as the encryption key is kept secret, it is impossible for the malicious user to recover the scrambling sequence, and thus disguised jamming can hardly be launched. In this case, the best jamming strategy for the malicious user would be distributing its total available power uniformly on the spread spectrum by randomly generating a PN sequence as the scrambling sequence.

As will be seen in Sect. 2.5, under the condition that the jammer has comparable power as the authorized user, the harm of this kind of jamming without knowing the secure scrambling sequence will actually become trivial.

2.4.3 Complexity Analysis

In this subsection, we compare the computational complexity of conventional scrambling and the secure scrambling. For this purpose, we only need to compare the complexity of the scrambling sequence generation methods.

For the secure scrambling, every 128-bit block of the scrambling sequence is generated through one AES encryption process. Here, using the conventional scrambling method as the baseline, we compare the number of instructions required by each method for every 128 bits and also the time required for every 128 bits using a Dell computer with 1024M RAM and 2.8 GHz CPU speed. The results

Table 2.1 Complexity comparison of conventional scrambling and secure scrambling. (© [2016] IEEE. Reprinted, with permission, from ref. [19])

	Number of operations required for every 128 bits				
Method	AND	OR	BIT-SHIFT	TOTAL	Time (in seconds)
Conv. scrambling	5376	5248	5376	16000	0.0226
Secure scrambling	7096	6644	8640	22380	0.0536

are provided in Table 2.1. As can be seen, the computational complexity of secure scrambling is comparable with that of the conventional scrambling process.

2.5 Capacity of CDMA with and without Secure Scrambling Under Disguised Jamming

Without secure scrambling, the jammer can launch disguised jamming toward the CDMA systems by exploiting the known code information and mimicking the authorized signal. In this case, it has been shown in Sect. 2.2.2 that the error probability of the symbol transmission is lower bounded by $\frac{M-1}{2M}$, where M is the constellation size. In this section, by applying the arbitrarily varying channel (AVC) model, we will show that due to the symmetricity between the authorized signal and jamming interference, the capacity of the traditional CDMA system (i.e., without secure scrambling) under worst disguised jamming is actually zero; on the other hand, with secure scrambling, the shared secure randomness between the transmitter and the receiver breaks the symmetricity between the authorized signal and jamming and hence ensures positive capacity under worst disguised jamming.

Recall that an AVC channel model is generally characterized using a kernel $W : \mathcal{S} \times \mathcal{J} \rightarrow \mathcal{Y}$, where $\mathcal{S}$ is the transmitted signal space, $\mathcal{J}$ is the jamming space (i.e., the jamming is viewed as the state of the arbitrarily varying channel), and $\mathcal{Y}$ is the estimated signal space. For any $\mathbf{s} \in \mathcal{S}$, $\mathbf{j} \in \mathcal{J}$, and $\mathbf{y} \in \mathcal{Y}$, $W(\mathbf{y}|\mathbf{s}, \mathbf{j})$ denotes the conditional probability that $\mathbf{y}$ is detected at the receiver, given that $\mathbf{s}$ is the transmitted signal and $\mathbf{j}$ the jamming.

Definition 2.1 ([12]) The AVC is said to have a symmetric kernel, if $\mathcal{S} = \mathcal{J}$ and $W(\mathbf{y}|\mathbf{s}, \mathbf{j}) = W(\mathbf{y}|\mathbf{j}, \mathbf{s})$ for any $\mathbf{s}, \mathbf{j} \in \mathcal{S}, \mathbf{y} \in \mathcal{Y}$.

Definition 2.2 ([12]) Define $\hat{W} : \mathcal{S} \times \mathcal{S} \rightarrow \mathcal{Y}$ by $\hat{W}(\mathbf{y}|\mathbf{s}, \mathbf{s}') \triangleq \sum_{\mathbf{j} \in \mathcal{J}'} \pi(\mathbf{j}|\mathbf{s}') W(\mathbf{y}|\mathbf{s}, \mathbf{j})$, where $\pi : \mathcal{S} \rightarrow \mathcal{J}'$ is a probability matrix and $\mathcal{J}' \subseteq \mathcal{J}$. If there exists a $\pi : \mathcal{S} \rightarrow \mathcal{J}'$ such that $\hat{W}(\mathbf{y}|\mathbf{s}, \mathbf{s}') = \hat{W}(\mathbf{y}|\mathbf{s}', \mathbf{s})$, $\forall \mathbf{s}, \mathbf{s}' \in \mathcal{S}$, $\forall \mathbf{y} \in \mathcal{Y}$, then W is said to be symmetrizable.

Concerning the capacity of the AVC channel, it was shown in [12] that *the deterministic code capacity*[5] *of an AVC for the average probability of error is positive if and only if the AVC is neither symmetric nor symmetrizable.*

2.5.1 Capacity of CDMA Systems without Secure Scrambling Under Disguised Jamming

Without secure scrambling, the codes employed by the authorized user can be regenerated by the jammer, and disguised jamming can thus be generated by applying the same codes but with a fake symbol. If we denote the fake symbol by $v \in \Omega$, in an isolated symbol period, the chip-rate disguised jamming can be represented as:

$$\mathbf{j} = v\mathbf{c} = [vc_0, vc_1, \ldots, vc_{N-1}], \tag{2.29}$$

where $\mathbf{c} = [c_0, c_1, \ldots, c_{N-1}]$ is the spreading code and $v \in \Omega$ the fake symbol. The authorized signal can similarly be written as:

$$\mathbf{s} = u\mathbf{c} = [uc_0, uc_1, \ldots, uc_{N-1}], \tag{2.30}$$

where $u \in \Omega$ is the authorized symbol. Taking both the noise and jamming into account, the received chip-rate signal can be written as:

$$\mathbf{r} = \mathbf{s} + \mathbf{j} + \mathbf{n}, \tag{2.31}$$

in which $\mathbf{n} = [n_0, n_1, \ldots, n_{N-1}]$ and $\mathbf{r} = [r_0, r_1, \ldots, r_{N-1}]$ denote the AWGN noise vector and received signal vector, respectively.

Define the authorized signal space as $\mathcal{S} = \{u\mathbf{c} | u \in \Omega\}$, where $\mathbf{c} = [c_0, c_1, \ldots, c_{N-1}]$ is the spreading code. It follows immediately that the disguised jamming space is:

$$\mathcal{J} = \{v\mathbf{c} | v \in \Omega\} = \mathcal{S}. \tag{2.32}$$

Let $\hat{u} \in \Omega$ be the estimated version of the authorized symbol "u" at the receiver and $W_O(\hat{u}|\mathbf{s}, \mathbf{j})$ the conditional probability that $\hat{u}$ is estimated given that the authorized signal is $\mathbf{s} \in \mathcal{S}$ and the disguised jamming is $\mathbf{j} \in \mathcal{S}$. Thus, the CDMA system under disguised jamming can be modeled as an AVC channel characterized by the probability matrix:

[5] A deterministic code capacity is defined by the capacity that can be achieved by a communication system, when it applies only one code pattern during the information transmission. In other words, the coding scheme is deterministic and can be readily repeated by other users [12].

$$W_0 : \mathcal{S} \times \mathcal{S} \rightarrow \Omega, \tag{2.33}$$

where W_0 is the kernel of the AVC.

As indicated in (2.32), the jamming and the authorized signal are fully symmetric as they are generated from exactly the same space $\mathcal{S}$. Note that the recovery of the authorized symbol is completely based on $\mathbf{r}$ in (2.31), so we further have:

$$W_0(\hat{u}|\mathbf{s}, \mathbf{j}) = W_0(\hat{u}|\mathbf{j}, \mathbf{s}). \tag{2.34}$$

Combining (2.32) and (2.34), and following Definition 2.1, we have the proposition below.

Proposition 2.1 *Under disguised jamming, the kernel of the AVC corresponding to a CDMA system without secure scrambling, W_0, is symmetric.*

The symmetricity of the AVC kernel explains why the error probability of the symbol transmission in CDMA systems without secure scrambling is lower bounded under disguised jamming, as indicated in (2.19). Applying the result in [12] that the deterministic code capacity of an AVC with a symmetric or symmetrizable kernel is zero, the proposition below follows immediately.

Proposition 2.2 *Under disguised jamming, the deterministic code capacity of a CDMA system without secure scrambling is zero.*

2.5.2 Symmetricity Analysis of CDMA Systems with Secure Scrambling Under Disguised Jamming

From the discussions above, it can be seen that disguised jamming is destructive to CDMA systems without secure scrambling, as zero capacity implies a complete failure in information transmission. In what follows, we will show how secure scrambling breaks the symmetricity between the authorized signal and jamming interference and evaluate the resulted performance gain in terms of error probability and capacity.

When the coding information of the authorized user is securely hidden from the jammer by the secure scrambling scheme, the best strategy for the jammer would be distributing its total available power uniformly over the entire spectrum, since CDMA systems are well known to be resistant to narrowband jamming. To this end, the jammer can spread its power by using a randomly generated spreading sequence. More specifically, if we define $\mathcal{D} = \{[d_0, d_1, \ldots, d_{N-1}] | d_n = \pm 1, \ \forall n\}$, and denote the randomly generated spreading sequence by $\mathbf{d} \in \mathcal{D}$, the chip-rate jamming sequence can be represented as:

$$\mathbf{j} = v\mathbf{d} = [vd_0, vd_1, \ldots, vd_{N-1}], \tag{2.35}$$

where $v \in \Omega$ is the fake symbol. The jamming space now becomes:

$$\mathcal{J} = \{v\mathbf{d} | v \in \Omega, \mathbf{d} \in \mathcal{D}\}. \tag{2.36}$$

We can see that without the coding information $\mathbf{c}$, the jamming, $\mathbf{j}$, can only be generated from a space much larger than the authorized signal space. More specifically, $\mathcal{J} \supset \mathcal{S}$. For any $\mathbf{j} \in \mathcal{J}$, the probability that $\mathbf{j} \in \mathcal{S}$ (i.e., the jamming falls into the authorized signal space by coincidentally repeating the authorized code $\mathbf{c}$ or its negative) is $\frac{1}{2^{N-1}}$, which approaches zero when N is reasonably large.

With the jamming space $\mathcal{J}$ as defined in (2.36), the AVC corresponding to the CDMA system with secure scrambling can be characterized by:

$$W : \mathcal{S} \times \mathcal{J} \rightarrow \Omega. \tag{2.37}$$

Based on the discussion above, $\mathcal{J} \neq \mathcal{S}$. That is, the jamming and the authorized signal are no longer symmetric. Following Definition 2.1, we have the proposition below.

Proposition 2.3 *Under disguised jamming, the kernel of the AVC corresponding to a CDMA system with secure scrambling, W, is nonsymmetric.*

Next, we will prove a stronger result: *W is actually nonsymmetrizable*. According to Definition 2.2, we need to show that for any probability matrix $\pi : \mathcal{S} \rightarrow \mathcal{J}$, there exists some $\mathbf{s}_0, \mathbf{s}'_0 \in \mathcal{S}$ and $\hat{u}_0 \in \Omega$, such that:

$$\hat{W}(\hat{u}_0|\mathbf{s}_0, \mathbf{s}'_0) \neq \hat{W}(\hat{u}_0|\mathbf{s}'_0, \mathbf{s}_0), \tag{2.38}$$

where $\hat{W}(\hat{u}|\mathbf{s}, \mathbf{s}') \triangleq \sum_{\mathbf{j} \in \mathcal{J}} \pi(\mathbf{j}|\mathbf{s}') W(\hat{u}|\mathbf{s}, \mathbf{j})$. To prove it, we present three lemmas first.

Lemma 2.1 *In a complex plane, there are two pairs of symmetric points, (u_1, u_2) and (v_1, v_2), which share the same axis of symmetry. Suppose u_1 and v_1 are located on one side of the axis of symmetry, while u_2 and v_2 reside on the other side. For any point p, if $|p - u_1| \leq |p - u_2|$, then $|p - v_1| \leq |p - v_2|$, where the equality holds if and only if $|p - u_1| = |p - u_2|$.*

Proof From $|p - u_1| \leq |p - u_2|$, we know that p is either on the same side with u_1 or exactly on the axis of symmetry. If p is on the same side with u_1, i.e., $|p - u_1| < |p - u_2|$, since u_1 and v_1 are on the same side, hence p and v_1 are on the same side. Since v_2 is on the other side, it follows immediately that $|p - v_1| < |p - v_2|$. If p is exactly on the axis of symmetry, i.e., $|p - u_1| = |p - u_2|$, then $|p - v_1| = |p - v_2|$. Similarly, if $|p - v_1| = |p - v_2|$, then $|p - u_1| = |p - u_2|$. ■

Define $R(u)$ as the region of detection for symbol $u \in \Omega$ in the complex plane, which means that any received symbol located in this region will be decided as "u" by a minimum distance detector. That is, for any point $p \in R(u)$, any symbol $v \in \Omega$ and $v \neq u$, we always have $|p - u| < |p - v|$. Furthermore, for a pair of symmetric

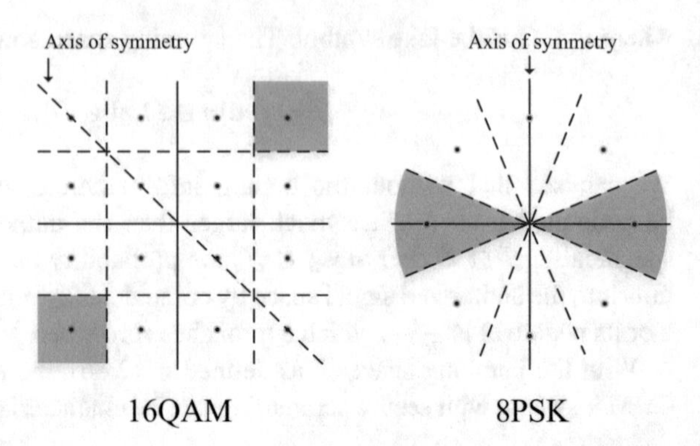

Fig. 2.2 Illustration of symmetric symbols with axial symmetric regions of detection. (© [2016] IEEE. Reprinted, with permission, from ref. [19])

symbols from a symmetric constellation,[6] $(u, -u)$, their regions of detection, $R(u)$ and $R(-u)$, are said to be axial symmetric, if, for any point $p \in R(u)$, there always exists a point $M(p) \in R(-u)$, such that $(p, M(p))$ and $(u, -u)$ share the same axis of symmetry. Such a point, $M(p)$, is called the *symmetric point* of p with respect to the axis of symmetry for $(u, -u)$. The shaded areas of Fig. 2.2 illustrate the regions of detection for two symmetric symbols, which are axial symmetric with respect to the axis of symmetry given in the figure.

Lemma 2.2 *Assume the received symbol is* $r = u + z + n$, *where* $u \in \Omega$ *is the transmitted symbol,* z *a fixed complex deviation with* $|z| \leq |u|$, *and* $n \sim \mathcal{CN}\left(0, \sigma^2\right)$ *the complex Gaussian noise. If the regions of detection,* $R(u)$ *and* $R(-u)$, *are axial symmetric, then we have* $W(u|u, z) \geq W(-u|u, z)$, *where the equality holds if and only if* $z = -u$.

Proof For the received symbol $r = u + z + n$, where "u" is the transmitted symbol, "z" is the fixed deviation, and $n \sim \mathcal{CN}\left(0, \sigma^2\right)$, r follows a complex Gaussian distribution, $r \sim \mathcal{CN}\left(u + z, \sigma^2\right)$. Hence, the conditional probability that the received symbol will be decided as "u" given that the actually transmitted symbol is "u" and the fixed deviation is "z" can be calculated as:

$$W(u|u, z) = \int_{r \in R(u)} f_R(r) dr, \tag{2.39}$$

where $f_R(r) = \frac{1}{\sqrt{2\pi}\sigma} \exp\left\{-\frac{|r-(u+z)|^2}{2\sigma^2}\right\}$ is the probability density function of r. Similarly, the probability that the received symbol will be decided as "$-u$" given that the actually transmitted symbol is "u" and the fixed deviation is "z" can be calculated as:

[6] A constellation Ω is said to be symmetric, if for any $u \in \Omega$, we always have $-u \in \Omega$. For maximum power efficiency, traditional constellations in use are generally symmetric, e.g., PSK and QAM.

$$W(-u|u,z) = \int_{r\in R(-u)} f_R(r)dr = \int_{r\in R(u)} f_R(M(r))dr. \tag{2.40}$$

Note that the two regions of detection, $R(u)$ and $R(-u)$, are axial symmetric, and $M(r)$ is the symmetric point of r with respect to the axis of symmetry for $(u, -u)$.

Let $p = u + z$, $u_1 = u$, and $u_2 = -u$. Since $|z| \le |u|$,

$$|p-u_1| - |p-u_2| = |z| - |2u+z| \le 0, \tag{2.41}$$

where the equality holds if and only if $z = -u$. For any $r \in R(u)$, r must be on the same side with $u_1 = u$ relative to the axis of symmetry for $(u, -u)$, and $M(r)$ must be on the same side with $u_2 = -u$, as illustrated in Fig. 2.2. Applying Lemma 2.1, it follows from (2.41) that:

$$|p-r| - |p-M(r)| = |r-(u+z)| - |M(r)-(u+z)| \le 0, \ \forall r \in R(u), \tag{2.42}$$

where the equality holds if and only if $z = -u$. Thus, we have:

$$\begin{aligned} &W(u|u,z) - W(-u|u,z) \\ &= \int_{r\in R(u)} [f_R(r) - f_R(M(r))]dr \\ &= \int_{r\in R(u)} \frac{1}{\sqrt{2\pi}\sigma}\left[\exp\left\{-\frac{|r-(u+z)|^2}{2\sigma^2}\right\} - \exp\left\{-\frac{|M(r)-(u+z)|^2}{2\sigma^2}\right\}\right] dr. \end{aligned} \tag{2.43}$$

Applying (2.42) to (2.43), we have:

$$W(u|u,z) - W(-u|u,z) \ge 0, \tag{2.44}$$

where the equality holds if and only if $z = -u$. ∎

Lemma 2.3 *Assume the received signal is $\mathbf{r} = \mathbf{s} + \mathbf{j} + \mathbf{n}$, where $\mathbf{s} = u\mathbf{c}$ is the signal vector with $u \in \Omega$ as the transmitted symbol and $\mathbf{c}$ as the spreading code, $\mathbf{j} \in \mathcal{J} = \{v\mathbf{d} | v \in \Omega, \mathbf{d} \in \mathcal{D}\}$ is the jamming vector, and $\mathbf{n}$ is the noise vector. If the regions of detection, $R(u)$ and $R(-u)$, are axial symmetric, and $|u| \ge |v|$, $\forall v \in \Omega$, then we have $W(u|\mathbf{s},\mathbf{j}) \ge W(-u|\mathbf{s},\mathbf{j})$, where the equality holds if and only if $\mathbf{j} = -\mathbf{s}$.*

Proof With $\mathbf{r} = \mathbf{s} + \mathbf{j} + \mathbf{n}$, the despread signal at the receiver would be:

$$r = \frac{1}{N}\sum_{n=0}^{N-1} r_n c_n = u + \frac{v}{N}\sum_{n=0}^{N-1} c_n d_n + \frac{1}{N}\sum_{n=0}^{N-1} c_n n_n. \tag{2.45}$$

Let $z = \frac{v}{N}\sum_{n=0}^{N-1} c_n d_n$ and $n = \frac{1}{N}\sum_{n=0}^{N-1} c_n n_n$. Note that for all n, $c_n = \pm 1$, so the despread noise n would follow a complex Gaussian distribution, i.e., $n \sim \mathcal{CN}(0, \frac{\sigma_n^2}{N})$, where σ_n^2 is the original noise power before despreading. Hence, the recovered symbol, $r = u + z + n$, is actually the transmitted symbol "u" distorted by a fixed deviation z and a complex Gaussian noise n.

Since $|v| \leq |u|$, and for all n, $c_n = \pm 1$, $d_n = \pm 1$, we know that $|z| = |\frac{v}{N}\sum_{n=0}^{N-1} c_n d_n| \leq |u|$. Applying Lemma 2.2, we have $W(u|u, z) \geq W(-u|u, z)$, where the equality holds if and only if $z = \frac{v}{N}\sum_{n=0}^{N-1} c_n d_n = -u$. We then prove that $z = -u$ is equivalent to $\mathbf{j} = -\mathbf{s}$. On one hand, if $z = -u$, then $|z| = |u|$. Considering $|v| \leq |u|$ and $|\frac{1}{N}\sum_{n=0}^{N-1} c_n d_n| \leq 1$, we must have $|v| = |u|$ and $|\frac{1}{N}\sum_{n=0}^{N-1} c_n d_n| = 1$. There are only two cases that satisfy $z = -u$: (1) $v = -u$ and $d_n = c_n$, $\forall n$; (2) $v = u$ and $d_n = -c_n$, $\forall n$. Both cases lead to $\mathbf{j} = v\mathbf{d} = -u\mathbf{c} = -\mathbf{s}$. On the other hand, if $\mathbf{j} = -\mathbf{s}$, then it leads to the same two cases as above, both of which satisfy $z = -u$.

Due to the equivalence between the signals before and after despreading as shown in (2.45), we have $W(u|u, z) = W(u|\mathbf{s}, \mathbf{j})$ and $W(-u|u, z) = W(-u|\mathbf{s}, \mathbf{j})$. It then follows immediately that $W(u|\mathbf{s}, \mathbf{j}) \geq W(-u|\mathbf{s}, \mathbf{j})$, where the equality holds if and only if $\mathbf{j} = -\mathbf{s}$. ■

Proposition 2.4 *Under disguised jamming, the kernel of the AVC corresponding to a CDMA system with secure scrambling, W, is nonsymmetrizable.*

Proof We will show that for any probability matrix $\pi : \mathcal{S} \to \mathcal{J}$, there exists some $\mathbf{s}_0, \mathbf{s}_0' \in \mathcal{S}$ and $\hat{u}_0 \in \Omega$, such that:

$$\hat{W}(\hat{u}_0|\mathbf{s}_0, \mathbf{s}_0') \neq \hat{W}(\hat{u}_0|\mathbf{s}_0', \mathbf{s}_0), \tag{2.46}$$

where $\hat{W}(\hat{u}|\mathbf{s}, \mathbf{s}') \triangleq \sum_{\mathbf{j}\in\mathcal{J}} \pi(\mathbf{j}|\mathbf{s}')W(\hat{u}|\mathbf{s}, \mathbf{j})$. To this end, we pick $\mathbf{s}_0 = u\mathbf{c}$, $\mathbf{s}_0' = -u\mathbf{c}$, and $\hat{u}_1 = u$ and $\hat{u}_2 = -u$. Note that "u" is picked such that $R(u)$ and $R(-u)$ are axial symmetric, and $|u| \geq |v|$, $\forall v \in \Omega$, as illustrated in Fig. 2.2. We will prove that $\hat{W}(\hat{u}_1|\mathbf{s}_0, \mathbf{s}_0') = \hat{W}(\hat{u}_1|\mathbf{s}_0', \mathbf{s}_0)$ and $\hat{W}(\hat{u}_2|\mathbf{s}_0, \mathbf{s}_0') = \hat{W}(\hat{u}_2|\mathbf{s}_0', \mathbf{s}_0)$ cannot hold simultaneously, by showing that:

$$\hat{W}(\hat{u}_1|\mathbf{s}_0, \mathbf{s}_0') - \hat{W}(\hat{u}_2|\mathbf{s}_0, \mathbf{s}_0') > \hat{W}(\hat{u}_1|\mathbf{s}_0', \mathbf{s}_0) - \hat{W}(\hat{u}_2|\mathbf{s}_0', \mathbf{s}_0). \tag{2.47}$$

Following the definition of $\hat{W}$, we have:

$$\begin{aligned}
&\hat{W}(\hat{u}_1|\mathbf{s}_0, \mathbf{s}_0') - \hat{W}(\hat{u}_2|\mathbf{s}_0, \mathbf{s}_0') \\
&= \sum_{\mathbf{j}\in\mathcal{J}} \pi(\mathbf{j}|\mathbf{s}_0')W(\hat{u}_1|\mathbf{s}_0, \mathbf{j}) - \sum_{\mathbf{j}\in\mathcal{J}} \pi(\mathbf{j}|\mathbf{s}_0')W(\hat{u}_2|\mathbf{s}_0, \mathbf{j}) \\
&= \sum_{\mathbf{j}\in\mathcal{J}} \pi(\mathbf{j}|\mathbf{s}_0')[W(\hat{u}_1|\mathbf{s}_0, \mathbf{j}) - W(\hat{u}_2|\mathbf{s}_0, \mathbf{j})].
\end{aligned} \tag{2.48}$$

Note that $W(\hat{u}_1|\mathbf{s}_0, \mathbf{j})$ and $W(\hat{u}_2|\mathbf{s}_0, \mathbf{j})$ denote the probabilities that the received symbol is decided as $\hat{u}_1 = u$ and $\hat{u}_2 = -u$, respectively, given that the transmitted signal is $\mathbf{s}_0$ and the jamming is $\mathbf{j}$. Applying Lemma 2.3, we have:

$$W(\hat{u}_1|\mathbf{s}_0, \mathbf{j}) \geq W(\hat{u}_2|\mathbf{s}_0, \mathbf{j}), \tag{2.49}$$

where the equality holds if and only if $\mathbf{j} = -\mathbf{s}_0$. Substituting (2.49) into (2.48), it follows immediately that:

$$\hat{W}(\hat{u}_1|\mathbf{s}_0, \mathbf{s}_0') - \hat{W}(\hat{u}_2|\mathbf{s}_0, \mathbf{s}_0') \geq 0, \tag{2.50}$$

where the equality holds if and only if $\pi(\mathbf{j}|\mathbf{s}_0') = 0, \ \forall \mathbf{j} \neq -\mathbf{s}_0$. This means that $\hat{W}(\hat{u}_1|\mathbf{s}_0, \mathbf{s}_0') = \hat{W}(\hat{u}_2|\mathbf{s}_0, \mathbf{s}_0')$ occurs only when the jammer can *always* generate the jamming exactly as the opposite to the authorized signal, which is impossible since the jammer has no knowledge how the spreading sequence $\mathbf{c}$ is encrypted and changes at each symbol period. Based on the observation above, we further have:

$$\hat{W}(\hat{u}_1|\mathbf{s}_0, \mathbf{s}_0') - \hat{W}(\hat{u}_2|\mathbf{s}_0, \mathbf{s}_0') > 0. \tag{2.51}$$

Applying the same methodology, we can show that:

$$\hat{W}(\hat{u}_1|\mathbf{s}_0', \mathbf{s}_0) - \hat{W}(\hat{u}_2|\mathbf{s}_0', \mathbf{s}_0) < 0. \tag{2.52}$$

Combining (2.51) and (2.52), we have:

$$\hat{W}(\hat{u}_1|\mathbf{s}_0, \mathbf{s}_0') - \hat{W}(\hat{u}_2|\mathbf{s}_0, \mathbf{s}_0') > \hat{W}(\hat{u}_1|\mathbf{s}_0', \mathbf{s}_0) - \hat{W}(\hat{u}_2|\mathbf{s}_0', \mathbf{s}_0), \tag{2.53}$$

which shows that $\hat{W}(\hat{u}_1|\mathbf{s}_0, \mathbf{s}_0') = \hat{W}(\hat{u}_1|\mathbf{s}_0', \mathbf{s}_0)$ and $\hat{W}(\hat{u}_2|\mathbf{s}_0, \mathbf{s}_0') = \hat{W}(\hat{u}_2|\mathbf{s}_0', \mathbf{s}_0)$ cannot hold simultaneously. ■

Since the kernel corresponding to a CDMA system with secure scrambling under disguised jamming, W, is neither symmetric (Proposition 2.3) nor symmetrizable (Proposition 2.4), we have the proposition below.

Proposition 2.5 *Under disguised jamming, the deterministic code capacity of a CDMA system with secure scrambling is not zero.*

Discussions Proposition 2.4 shows that the kernel of the AVC corresponding to a CDMA system with secure scrambling is nonsymmetrizable, except when the jammer can always generate the jamming as exactly as the negative of the authorized signal. However, this is computationally impossible, since it is equivalent to break AES applied in secure scrambling, which has been proved to be secure under all known attacks.

An aggressive jammer can probably launch jamming consisting of multiple spreading codes, in order to increase the probability that one of its applied codes coincides with the one applied by the authorized user. When the number of spreading codes covered by the jammer is small, the harm to the authorized communication would be negligible. While using multiple spreading codes produces more effective jamming, the power consumption can be forbiddingly high. However, it does indicate that when the user information (including both symbol and codes) is unknown, the most effective jamming is still Gaussian, resulting from accumulation of a large number of spreading codes and the central limit theorem (CLT).

2.5.3 *Capacity Calculation of CDMA Systems with Secure Scrambling Under Disguised Jamming*

So far we have shown that in CDMA systems with secure scrambling, the symmetricity between the authorized signal and the disguised jamming is broken, and hence the capacity is no longer zero. A natural question is: what is the capacity then? Although it is difficult to derive a modulation-specific capacity, we manage to provide a general analysis on the capacity by applying the Shannon formula as stated below. For particular modulation schemes like QAM and PSK, the error probabilities of symbol transmission will also be provided.

Recall that at the receiver, the despread symbol under disguised jamming can be calculated as:

$$r = \frac{1}{N}\sum_{n=0}^{N-1} r_n c_n = u + \frac{v}{N}\sum_{n=0}^{N-1} c_n d_n + \frac{1}{N}\sum_{n=0}^{N-1} c_n n_n. \tag{2.54}$$

Note that for all n, $c_n = \pm 1$ are constant, while $d_n = \pm 1$ are statistically independent and identically distributed (i.i.d.) binary random variables with zero mean and variance 1. Applying the central limit theorem (CLT), $\frac{1}{N}\sum_{n=0}^{N-1} c_n d_n$ would follow a complex Gaussian distribution with zero mean and variance $\frac{1}{N}$, i.e.:

$$\frac{1}{N}\sum_{n=0}^{N-1} c_n d_n \sim \mathcal{CN}\left(0, \frac{1}{N}\right). \tag{2.55}$$

Similarly, we have:

$$\frac{1}{N}\sum_{n=0}^{N-1} c_n n_n \sim \mathcal{CN}\left(0, \frac{\sigma_n^2}{N}\right), \tag{2.56}$$

where σ_n^2 is the original noise power before despreading. It then follows that r is also a complex Gaussian variable, whose distribution can be characterized by:

$$r \sim \mathcal{CN}\left(u, \frac{|v|^2}{N} + \frac{\sigma_n^2}{N}\right), \tag{2.57}$$

which implies that for an arbitrary transmitted symbol $u \in \Omega$ and an arbitrary fake symbol $v \in \Omega$ in (2.54), the received symbol is actually the transmitted symbol "u" polluted by a complex Gaussian noise, $n \sim \mathcal{CN}\left(0, \frac{|v|^2}{N} + \frac{\sigma_n^2}{N}\right)$.

Let σ_s^2 denote the average symbol power, namely, $\mathcal{E}\{|u|^2\} = \sigma_s^2$, where $u \in \Omega$. Based on (2.57), for a specific fake symbol $v \in \Omega$, the corresponding signal-to-jamming-and-noise ratio (SJNR) can be calculated as:

$$\gamma(v) = \frac{\sigma_s^2}{|v|^2/N + \sigma_n^2/N} = \frac{N\sigma_s^2}{|v|^2 + \sigma_n^2}. \tag{2.58}$$

The symbol error probability largely depends on the employed constellation Ω. However, with SJNR available, and considering all possible $v \in \Omega$, the average symbol error probability can be calculated as:

$$\mathcal{P}_s = \frac{1}{|\Omega|} \sum_{v \in \Omega} \mathcal{P}_\Omega(\gamma(v)) = \frac{1}{|\Omega|} \sum_{v \in \Omega} \mathcal{P}_\Omega\left(\frac{N\sigma_s^2}{|v|^2 + \sigma_n^2}\right), \tag{2.59}$$

where $|\Omega|$ denotes the constellation size, and $\mathcal{P}_\Omega(\cdot)$ is readily available in [21, eqn. (5.2-78) & (5.2-79), page 278] for QAM and [21, eqn. (5.2-56), page 268] for PSK, respectively.

To calculate the capacity, a CDMA system which operates over a spectrum of B Hz can be equivalently viewed as a narrowband transmission with a bandwidth of $\frac{B}{N}$, while simultaneously having its SJNR level increased to (2.58) as a result of the processing gain. Hence, the capacity can be obtained as:

$$C = \frac{B}{N} \frac{1}{|\Omega|} \sum_{v \in \Omega} \log_2(1 + \gamma(v)) = \frac{B}{N} \frac{1}{|\Omega|} \sum_{v \in \Omega} \log_2\left(1 + \frac{N\sigma_s^2}{|v|^2 + \sigma_n^2}\right). \tag{2.60}$$

For clarity, we summarize the analysis above in Table 2.2. It can be seen that (1) the symbol error probability of a CDMA system under disguised jamming can be decreased significantly using the secure scrambling scheme, compared with the lower-bounded error probability without secure scrambling, especially when the processing gain, N, is large, and (2) with secure scrambling, the capacity of a CDMA system will no longer be zero.

Overall, we would like to point out that based on the shared secret between the authorized transmitter and receiver, secure scrambling enhances the randomness in the CDMA spreading process and makes it forbiddingly difficult for the malicious user to launch disguised jamming. Our results echo the observations in [8, 12, 22–24], where random coding is viewed as a promising solution in combating disguised jamming.

Table 2.2 Comparison of CDMA systems with and without secure scrambling under disguised jamming. (© [2016] IEEE. Reprinted, with permission, from ref. [19])

	Without secure scrambling	With secure scrambling
Symmetric	Yes	No
Symmetrizable	N/A	No
SJNR	N/A	$\frac{N\sigma_s^2}{\lvert v\rvert^2+\sigma_n^2},\ v \in \Omega$
Error probability	$\geq \frac{M-1}{2M}$	$\frac{1}{\lvert\Omega\rvert}\sum_{v\in\Omega}\mathcal{P}_\Omega\left(\frac{N\sigma_s^2}{\lvert v\rvert^2+\sigma_n^2}\right)$
Capacity	0	$\frac{B}{N}\frac{1}{\lvert\Omega\rvert}\sum_{v\in\Omega}\log_2\left(1+\frac{N\sigma_s^2}{\lvert v\rvert^2+\sigma_n^2}\right)$

2.6 Numerical Results

In this section, we numerically evaluate the effectiveness of the presented jamming mitigation schemes: robust receiver design and secure scrambling. In what follows, we assume AWGN channels and launch two separate simulation settings from practical CDMA systems, which, we believe, provide good examples in potential applications of those schemes.

2.6.1 *Jamming Mitigation with Robust Receiver Design*

In this subsection, through several simulation examples, we first evaluate the performance degradation of CDMA systems under disguised jamming and then demonstrate the effectiveness of the designed receiver in jamming estimation and BER performance improvement. In the simulation, we adopt the settings as in civilian GPS, where BPSK modulation is applied and the spreading code is a Gold sequence with a processing gain $N = 1023$. Note that the civilian GPS has public spreading codes, and it is exactly one of the scenarios where the robust receiver design is needed in order to avoid the code concealment. Moreover, we set the oversampling factor to 32, which means that there are 32 samples in each chip with a T_c duration. Note that the oversampling factor determines the resolution of the timing difference estimation, i.e., $\frac{1}{32}T_c$, for the current setting.

(1) Performance Degradation of Conventional CDMA Systems Under Disguised Jamming In this simulation example, we evaluate the impact of disguised jamming with different timing differences on the BER performance of the conventional CDMA system. The amplitude ratio γ is set to 1, and we apply the conventional CDMA receiver as in (2.14) without any jamming estimation. It is observed from Fig. 2.3 that comparing with jamming-free case, the BER performance is severely degraded by the disguised jamming, especially when

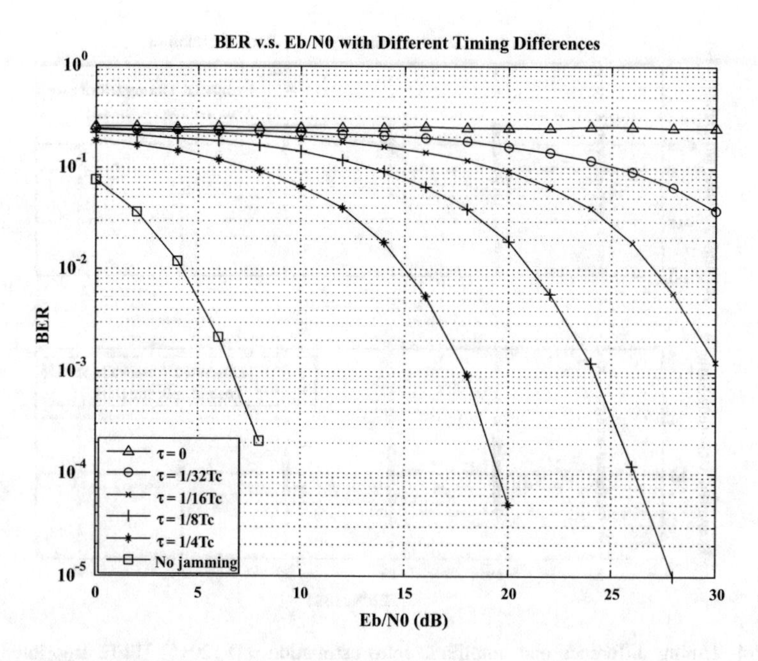

Fig. 2.3 BER v.s. Eb/N0 for the conventional CDMA receiver under various disguised jamming. (© [2016] IEEE. Reprinted, with permission, from ref. [19])

the timing difference τ is small. In the worst case with $\tau = 0$, the BER maintains at approximately $\frac{1}{4}$ no matter how high the SNR is, which agrees with the lower bound in (2.19).

(2) Timing Difference and Amplitude Ratio Estimation In this simulation example, we provide the estimation results of the timing difference τ and amplitude ratio γ by applying the robust CDMA receiver. Here we set $\tau = \frac{1}{4}T_c$ and $\gamma = 1.2$. In Fig. 2.4, we can observe that both the timing difference and amplitude ratio can be accurately estimated with reasonable SNRs, and the accuracy improves as the SNR increases.

(3) BER Performance Improvement with Jamming Estimation In this simulation example, we compare the BER performance of the robust CDMA receiver with that of the conventional receiver. To explore a time-varying jamming scenario, the timing difference τ is set to be uniformly distributed on $[-\frac{1}{4}T_c, 0) \cup (0, \frac{1}{4}T_c]$, and the amplitude ratio γ follows a normal distribution $\mathcal{N}(1, \sigma^2)$, where $\sigma = \frac{1}{6}$. Note that we do not take into account $\tau = 0$, in which case the BER cannot be decreased because of the lower bound in (2.19). In Fig. 2.5, it is observed that the BER is decreased significantly by the robust CDMA receiver with reasonable SNRs. With low SNRs, the BER cannot be decreased due to the inaccurate jamming estimation, which demonstrates that it is more difficult to combat disguised jamming under poor channel conditions.

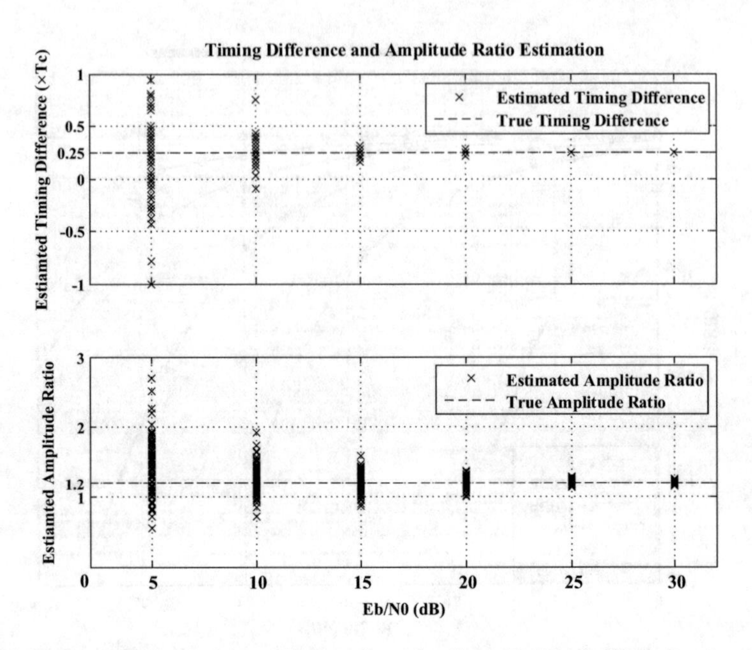

Fig. 2.4 Timing difference and amplitude ratio estimation. (© [2016] IEEE. Reprinted, with permission, from ref. [19])

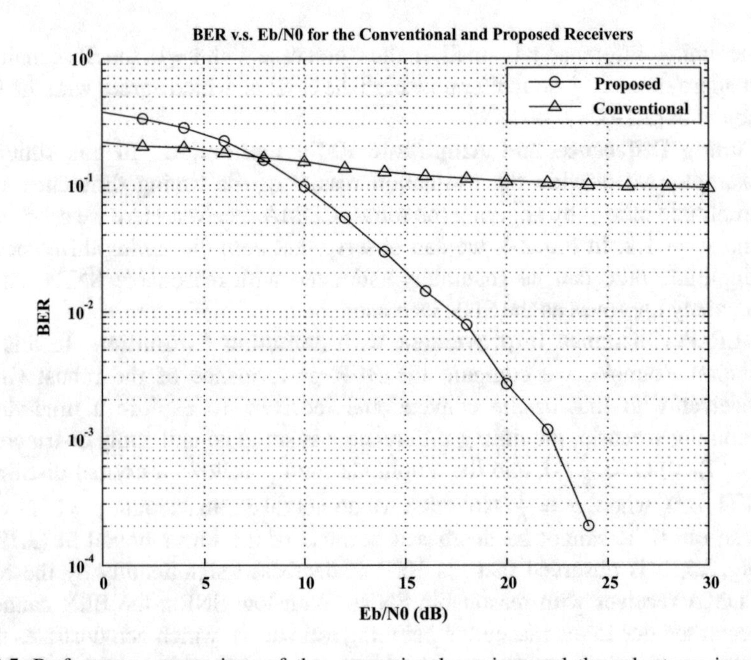

Fig. 2.5 Performance comparison of the conventional receiver and the robust receiver under disguised jamming. (© [2016] IEEE. Reprinted, with permission, from ref. [19])

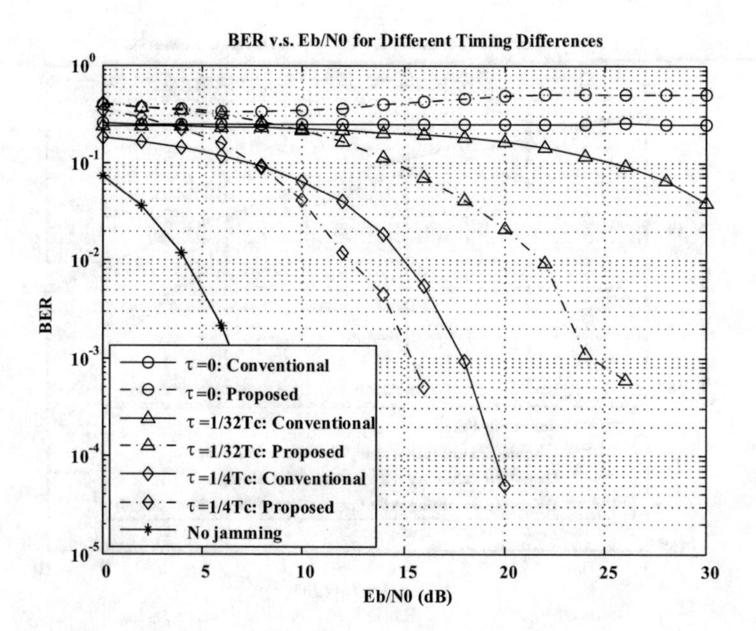

Fig. 2.6 BER v.s. Eb/N0 for different timing differences. (© [2016] IEEE. Reprinted, with permission, from ref. [19])

To evaluate how well the robust receiver works with different but fixed timing differences, we compare the performance of the conventional receiver with that of the robust receiver regrading different timing differences in Fig. 2.6, where the amplitude ratio γ is set to 1. It is observed that (i) for nonzero timing differences, the BER is decreased significantly by the robust CDMA receiver with reasonable SNRs, and (ii) for the worst disguised jamming with zero timing difference, the robust receiver design cannot help at all, in which case we should consider using secure scrambling to break the symmetricity.

2.6.2 *Jamming Mitigation with Secure Scrambling*

In this subsection, we numerically show the effectiveness of the secure scrambling in combating disguised jamming for CDMA systems whose scrambling codes can potentially be protected. In the simulation, we adopt Walsh codes with a processing gain $N = 64$ as the spreading codes, and apply 16QAM modulation. The symbol error rates (SERs) of CDMA systems are shown in Fig. 2.7 associating with the following four conditions: (a) jamming-free case as the benchmark, (b) under disguised jamming but without secure scrambling, (c) under disguised jamming and with secure scrambling, and (d) the theoretical result for the case in (c) as a verification.

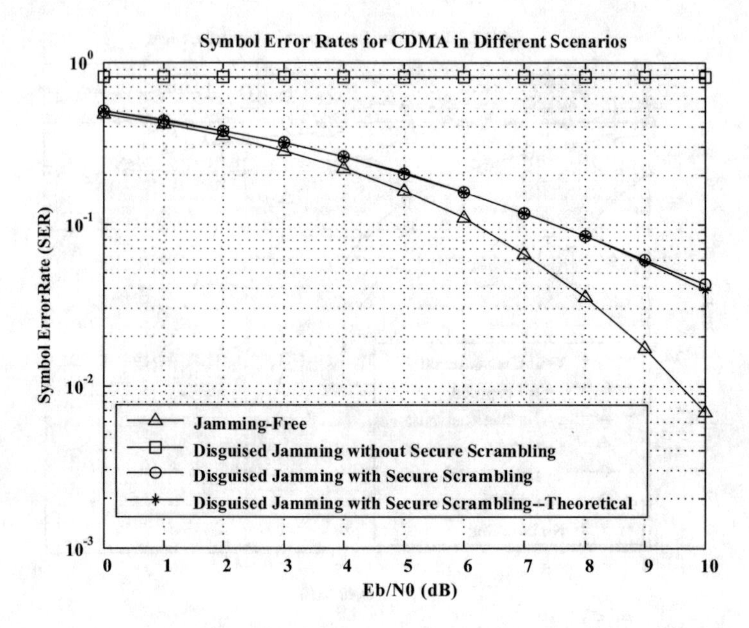

Fig. 2.7 Symbol error rates (SERs) for CDMA in different scenarios. (© [2016] IEEE. Reprinted, with permission, from ref. [19])

In Fig. 2.7, it is observed that (1) without secure scrambling, the symbol error rate of CDMA communication under disguised jamming maintains at an extremely high level no matter how high the SNR is, which shows that the CDMA communication is severely paralyzed by disguised jamming; (2) the secure scrambling scheme significantly improves the performance of CDMA communication under disguised jamming, where the SER curve matches the theoretical result as indicated in (2.59) as well; and (3) the SER curve using secure scrambling under disguised jamming is quite close to that of the jamming-free case, and it can be expected that the gap will become even smaller if we have a larger processing gain N.

2.7 Conclusions

In this chapter, we analyzed the impact of disguised jamming on conventional CDMA systems and developed two effective approaches to mitigate the jamming effect for two different categories of CDMA systems. *For CDMA systems with public codes which cannot be concealed for some reason*, we mitigated the disguised jamming through robust receiver design. The presented approach exploited the small timing difference between the authorized signal and the jamming interference. We estimated the authorized symbols as well as the jamming parameters by finding the minimum mean square error (MMSE) between the received signal and jammed signal, which is the sum of the authorized signal and the disguised jamming. The numerical results demonstrated that with reasonable SNRs, the robust receiver significantly improves the BER performance of CDMA systems under disguised jamming and also provides a good evaluation about jamming. *For CDMA systems which allow code concealment*, we mitigated disguised jamming using secure scrambling. Instead of using conventional scrambling codes, we applied Advanced Encryption Standard (AES) to generate the security-enhanced scrambling codes. Theoretical analysis shows that the capacity of the conventional CDMA systems without secure scrambling under disguised jamming is actually zero; however, the capacity can be significantly increased when the CDMA systems are protected using secure scrambling. Numerical examples were provided to demonstrate the effectiveness of secure scrambling in combating disguised jamming.

References

1. S. Barbarossa and A. Scaglione, "Adaptive time-varying cancellation of wideband interferences in spread-spectrum communications based on time-frequency distributions," *IEEE Transactions on Signal Processing*, vol. 47, no. 4, pp. 957–965, Apr 1999.
2. S. Aromaa, P. Henttu, and M. Juntti, "Transform-selective interference suppression algorithm for spread-spectrum communications," *IEEE Signal Processing Letters*, vol. 12, no. 1, pp. 49–51, Jan 2005.
3. C.-L. Wang and K.-M. Wu, "A new narrowband interference suppression scheme for spread-spectrum CDMA communications," *IEEE Transactions on Signal Processing*, vol. 49, no. 11, pp. 2832–2838, Nov 2001.
4. H. Holma and A. Toskala, *WCDMA for UMTS: HSPA Evolution and LTE*. New York, NY, USA: John Wiley & Sons, Inc., 2007.
5. E. Kaplan, *Understanding GPS - Principles and applications*, 2nd ed. Artech House, December 2005.
6. J. Massey, "Shift-register synthesis and BCH decoding," *IEEE Transactions on Information Theory*, vol. 15, no. 1, pp. 122–127, Jan 1969.
7. T. Li, Q. Ling, and J. Ren, "Physical layer built-in security analysis and enhancement algorithms for CDMA systems," *EURASIP Journal on Wireless Communications and Networking*, vol. 2007, no. 1, p. 083589, 2007.
8. T. Ericson, "The noncooperative binary adder channel," *IEEE Transactions on Information Theory*, vol. 32, no. 3, pp. 365–374, 1986.
9. M. Medard, "Capacity of correlated jamming channels," in *Allerton Conference on Communications, Computing and Control*, 1997.
10. L. Zhang, H. Wang, and T. Li, "Anti-jamming message-driven frequency hopping-part i: System design," *IEEE Transactions on Wireless Communications*, vol. 12, no. 1, pp. 70–79, Jan. 2013.
11. R. Nikjah and N. C. Beaulieu, "On antijamming in general cdma systems-part i: multiuser capacity analysis," *IEEE Transactions on Wireless Communications*, vol. 7, no. 5, pp. 1646–1655, May 2008.
12. T. Ericson, "Exponential error bounds for random codes in the arbitrarily varying channel," *IEEE Transactions on Information Theory*, vol. 31, no. 1, pp. 42–48, 1985.
13. S. Bhashyam and B. Aazhang, "Multiuser channel estimation and tracking for long-code CDMA systems," *IEEE Transactions on Communications*, vol. 50, no. 7, pp. 1081–1090, Jul 2002.

14. C. J. Escudero, U. Mitra, and D. T. M. Slock, "A Toeplitz displacement method for blind multipath estimation for long code DS/CDMA signals," *IEEE Transactions on Signal Processing*, vol. 49, no. 3, pp. 654–665, Mar 2001.
15. A. J. Weiss and B. Friedlander, "Channel estimation for DS-CDMA downlink with aperiodic spreading codes," *IEEE Transactions on Communications*, vol. 47, no. 10, pp. 1561–1569, Oct 1999.
16. M. Sahmoudi and M. Amin, "Fast iterative maximum-likelihood algorithm (FIMLA) for multipath mitigation in the next generation of GNSS receivers," *IEEE Transactions on Wireless Communications*, vol. 7, no. 11, pp. 4362–4374, November 2008.
17. Z. Xu and P. Liu, "Code-constrained blind detection of CDMA signals in multipath channels," *IEEE Signal Processing Letters*, vol. 9, no. 12, pp. 389–392, Dec 2002.
18. *Advanced Encryption Standard*, ser. FIPS-197, National Institute of Standards and Technology Std., Nov. 2001. [Online]. Available: http://csrc.nist.gov/publications/fips/fips197/fips-197.pdf
19. T. Song, K. Zhou, and T. Li, "CDMA system design and capacity analysis under disguised jamming," *IEEE Transactions on Information Forensics and Security*, vol. 11, no. 11, pp. 2487–2498, Nov 2016.
20. *Data Encryption Standard*, ser. FIPS-46-3, National Institute of Standards and Technology Std., Oct. 1999. [Online]. Available: http://csrc.nist.gov/publications/fips/fips46-3/fips46-3.pdf
21. J. G. Proakis, *Digital Communications*, 4th ed. New York: McGraw-Hill, 2000.
22. D. Blackwell, L. Breiman, and A. J. Thomasian, "The capacities of certain channel classes under random coding," *The Annals of Mathematical Statistics*, vol. 31, no. 3, pp. 558–567, 1960.
23. R. Ahlswede, "Elimination of correlation in random codes for arbitrarily varying channels," *Z. Wahrscheinlichkeitstheorie und Verwandte Gebiete*, vol. 44, no. 2, pp. 159–175, 1978.
24. A. Lapidoth and P. Narayan, "Reliable communication under channel uncertainty," *IEEE Transactions on Information Theory*, vol. 44, no. 6, pp. 2148–2177, 1998.

Chapter 3
Message-Driven Frequency Hopping Systems

3.1 The Concept of Message-Driven Frequency Hopping (MDFH)

3.1.1 *A Brief Revisit to Existing Work*

Recall that in traditional FH systems, the transmitter hops in a pseudo-random manner among the available frequencies according to a pre-specified algorithm, the receiver, and then operates in strict synchronization with the transmitter and remains tuned to the same center frequency. FH systems aim to minimize the possibility of hostile jamming and unauthorized interception through pseudo-random frequency hopping during the radio transmission.

In 1978, Cooper and Nettleton [1] first proposed a frequency hopping multiple access (FHMA) system with differential phase shift-keyed (DPSK) signaling for mobile communications. Later in the same year, Viterbi [2] initiated the use of multiple frequency shift keying (MFSK) for low-rate multiple access mobile satellite systems. After that, MFSK has been widely adopted in FHMA systems [3–6]. There are two major limitations with the conventional FH scheme: (i) *Strong requirement on frequency acquisition.* In existing FH systems, exact frequency synchronization has to be kept between the transmitter and the receiver. The strict requirement on synchronization directly influences the complexity, design, and performance of the system [7] and turns out to be a significant challenge in fast hopping system design. (ii) *Low spectral efficiency over large bandwidth.* Typically, FH systems require large bandwidth, which is proportional to the hopping rate and the number of all the available channels. In conventional FHMA, each user hops independently based on its own PN sequence, a collision occurs whenever there are two users over the same frequency band. Mainly limited by the collision effect, the spectral efficiency of conventional FH systems is very low. In literature, considerable efforts have been devoted to increasing the spectral efficiency of FH systems by applying high-dimensional modulation schemes [8–14]. Along with the ever-increasing demand

T. Li et al., *Wireless Communications Under Hostile Jamming: Security and Efficiency*, https://doi.org/10.1007/978-981-13-0821-5_3

on inherently secure high data-rate wireless communications, new techniques that are more efficient and reliable need to be developed.

3.1.2 The Basic Idea of MDFH

Message-driven frequency hopping (MDFH) is a three-dimensional modulation scheme [15]. The basic idea of MDFH is that part of the message acts as the PN sequence for carrier frequency selection at the transmitter. More specifically, selection of carrier frequencies is directly controlled by the encrypted information stream rather than by a pre-selected pseudo-random sequence as in conventional FH.

At the receiver, the transmitting frequency is captured using a filter bank as in the FSK receiver rather than using the frequency synthesizer. As a result, the carrier frequency (hence the information embedded in frequency selection) can be blindly detected at each hop. This relaxes the burden of strict frequency synchronization at the receiver. In addition, MDFH[1] makes it possible for faster frequency hopping in wideband systems.

The most significant property of MDFH is that by embedding a large portion of information into the hopping frequency selection process, additional information transmission is achieved with no extra cost on either bandwidth or power. In fact, transmission through hopping frequency control adds another dimension to the signal space, and the resulted coding gain can increase the system spectral efficiency by multiple times.

3.1.3 Transmitter Design

Let N_c be the total number of available channels, with $\{f_1, f_2, \cdots, f_{N_c}\}$ being the set of all available carrier frequencies. Ideally, all the available channels should be involved in the hopping frequency selection process, as is required by current frequency hopping specifications (e.g., Bluetooth). The number of bits used here to specify an individual channel is $B_c = \lfloor \log_2 N_c \rfloor$, where $\lfloor x \rfloor$ denotes that largest integer less than or equal to x. If N_c is a power of 2, then there exists a 1–1 map between the B_c-bit strings and the total available channels; otherwise, when N_c is not a power of 2, we will allow some B_c-bit strings to be mapped to more than one channel. More specifically, for $i = 1, \cdots, N_c$, the ith channel will be associated with the binary representation of the modulated channel index, $[(i-1) \mod 2^{B_c}] + 1$. In the following, for simplicity of notation, we assume that $N_c = 2^{B_c}$.

[1]It is interesting to note that in [16], the message is used to select the spreading code in code-division multiple access, which is the counterpart of FH in spread-spectrum techniques.

Let Ω be the selected constellation that contains M symbols; each symbol in the constellation represents $B_s = \log_2 M$ bits. Let T_s and T_h denote the symbol period and the hop duration, respectively, and then the number of hops per symbol period is given by $N_h = \frac{T_s}{T_h}$. We assume that N_h is an integer larger than or equal to 1. In other words, we focus on fast hopping systems.

We start by dividing the data stream into blocks of length $L \triangleq N_h B_c + B_s$. Each block is parsed into $N_h B_c$ *carrier bits* and B_s *ordinary bits*. The *carrier bits* are used to determine the hopping frequencies, and the *ordinary bits* are mapped to a symbol which is transmitted through the selected channels successively. Denote the nth block by X_n. The carrier bits in block X_n are further grouped into N_h vectors of length B_c, denoted by $[X_{n,1}, \cdots, X_{n,N_h}]$. The bit vector composed of B_s ordinary bits is denoted by Y_n, as shown in Fig. 3.1. We will transmit X_n within one symbol period.

The transmitter block diagram of the MDFH scheme is illustrated in Fig. 3.2. Each input data block, X_n, is fed into a serial-to-parallel (S/P) converter, where the carrier bits and the ordinary bits are split into two parallel data streams. The selected carrier frequencies corresponding to the nth block are denoted by $\{f_{n,1}, \cdots, f_{n,N_h}\}$, where each $f_{n,i} \in \{f_1, f_2, \cdots, f_{N_c}\}, \forall i \in \{1, \cdots, N_h\}$. Assume Y_n is mapped to symbol A_n, and the corresponding baseband signal is denoted as $m(t)$.

If PAM (pulse amplitude modulation) is adopted for baseband signal generation, then:

$$m(t) = \sum_{n=-\infty}^{\infty} \sum_{i=1}^{N_h} A_n \, g(t - nT_s - (i-1)T_h), \tag{3.1}$$

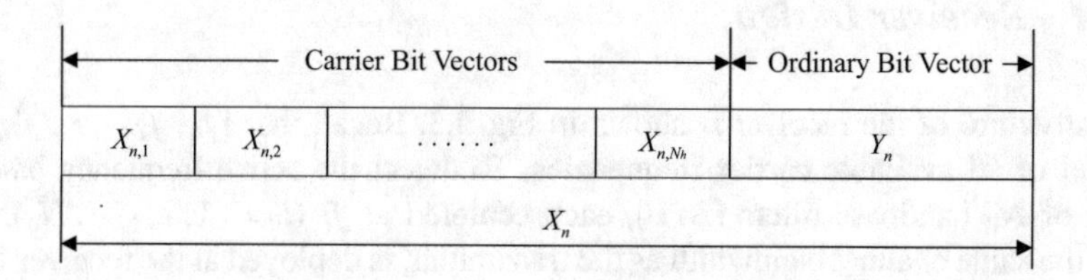

Fig. 3.1 The nth block of the information data. (© [2009] IEEE. Reprinted, with permission, from Ref. [15])

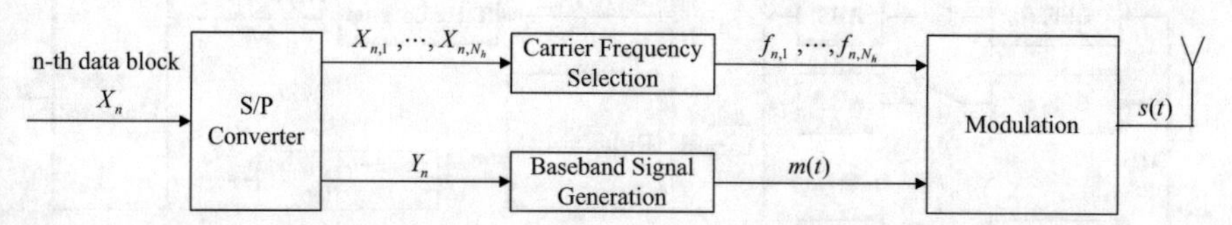

Fig. 3.2 Transmitter structure of MDFH. (© [2009] IEEE. Reprinted, with permission, from Ref. [15])

where $g(t)$ is the pulse shaping filter. Define $m_{n,i}(t) \triangleq A_n\ g(t - nT_s - (i-1)T_h)$, then $m(t) = \sum_{n=-\infty}^{\infty} \sum_{i=1}^{N_h} m_{n,i}(t)$. The corresponding passband waveform can be obtained as:

$$s(t) = \sqrt{\frac{2}{T_h}} Re\{ \sum_{n=-\infty}^{\infty} \sum_{i=1}^{N_h} m_{n,i}(t) e^{j2\pi f_{n,i} t} \chi_{n,i}(t) \}, \tag{3.2}$$

where:

$$\chi_{n,i}(t) = \begin{cases} 1, \ t \in (nT_s + (i-1)T_h, nT_s + iT_h], \\ 0, \ \text{otherwise}. \end{cases} \tag{3.3}$$

If MFSK is utilized for baseband modulation, then:

$$s(t) = \sqrt{\frac{2}{T_h}} \sum_{n=-\infty}^{\infty} \sum_{i=1}^{N_h} \cos 2\pi [f_{n,i} t + K_f \int_{-\infty}^{t} m_{n,i}(\tau) d\tau] \chi_{n,i}(t). \tag{3.4}$$

where K_f is a pre-selected constant.

Since FM generally requires much larger bandwidth than PAM, here we will mainly consider PAM-based FH systems, for which the overall bandwidth request is proportional to $\frac{N_c}{T_h}$.

3.1.4 Receiver Design

The structure of the receiver is shown in Fig. 3.3. Recall that $\{f_1, f_2, \cdots, f_{N_c}\}$ is the set of all available carrier frequencies. To detect the active frequency band, a bank of N_c bandpass filters (BPF), each centered at f_i $(i = 1, 2, \cdots, N_c)$, and with the same channel bandwidth as the transmitter, is deployed at the receiver front

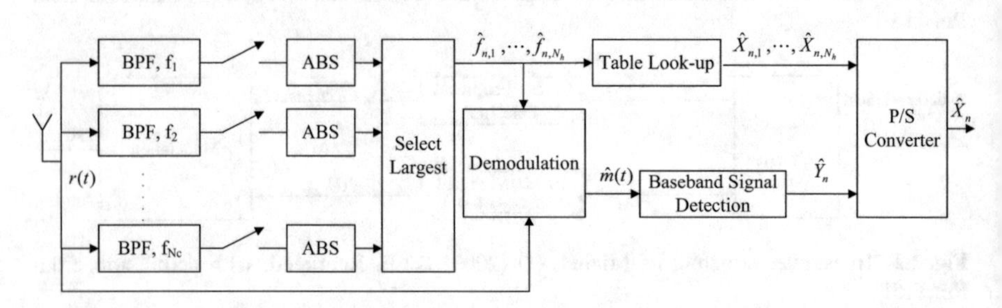

Fig. 3.3 Receiver structure of MDFH, here ABS means taking the absolute value. (© [2009] IEEE. Reprinted, with permission, from Ref. [15])

end. Since only one frequency band is occupied at any given moment, we simply measure the outputs of bandpass filters at each possible signaling frequency. The actual carrier frequency at a certain hopping period can be detected by selecting the one that captures the strongest signal. As a result, blind detection of the carrier frequency is achieved at the receiver.

More specifically, the received signal can be written as:

$$r(t) = h(t) * s(t) + w(t), \tag{3.5}$$

where $*$ stands for convolution, $h(t)$ is the channel impulse response, and $w(t)$ denotes additive white Gaussian noise. Accordingly, the outputs of bandpass filters are given by:

$$z_i(t) = q_i(t) * r(t), \quad \text{for } i = 1, \cdots, N_c, \tag{3.6}$$

where $q_i(t)$ is the ideal bandpass filter centered at frequency f_i.

If the channel is ideal, i.e., $h(t) = \delta(t)$, then:

$$z_i(t) = q_i(t) * s(t) + u_i(t), \quad \text{for } i = 1, \cdots, N_c, \tag{3.7}$$

where $u_i(t) = q_i(t) * w(t)$ is the filtered version of the noise. If the signal-to-noise ratio is sufficiently high, as in most communication systems, there is one and only one significantly strong signal among the outputs of the filter bank. Suppose that the lth filter captures this distinctive signal during the ith hop of the nth block, such that $\hat{f}_{n,i} = f_l$. The same procedures can be carried out to determine the carrier frequency at each hop.

Next, the estimated hopping frequencies $\{\hat{f}_{n,1}, \cdots, \hat{f}_{n,N_h}\}$ are used to extract the input signal. $\{\hat{f}_{n,1}, \cdots, \hat{f}_{n,N_h}\}$ are mapped back to B_c-bit strings to recover the carrier bits. We denote the estimated carrier bit vectors as $\{\hat{X}_{n,1}, \cdots, \hat{X}_{n,N_h}\}$. The ordinary bit vector, Y_n, is first estimated independently for each hop, and then bit-wise majority voting is applied for all the N_h estimates to make the final decision on each ordinary bit in Y_n. We denote the estimated ordinary bit vector as $\hat{Y}_n$. It then follows that the estimate of the nth block X_n can be obtained as: $\hat{X}_n = [\hat{X}_{n,1}, \cdots, \hat{X}_{n,N_h}, \hat{Y}_n]$. Below are some further discussions on MDFH:

- Unlike the conventional FH scheme for which the receiver can be designed as a single frequency synthesizer, the receiver that we use in MDFH is a filter bank consisting of bandpass filters, which is similar to that used for incoherent detection of FSK (frequency shift keying) signals. In fact, it can be implemented by putting several FSK receivers in parallel. Along with advances in large-scale circuit integration and fabrication, this receiver design is both feasible and practical.
- The design of the MDFH receiver leads to an observation regarding security: if such a filter bank is available to a malicious user, then all FH signals can largely

be intercepted by an unauthorized party. This implies that to prevent unauthorized interception, information has to be encrypted before being transmitted over an FH system.

- One important feature of FH systems is jamming resistance. As shown in Sect. 3.3.3, while achieving higher spectral efficiency, MDFH is also much more robust than the conventional FH under strong jamming interference. This is because that the performance of the conventional FH is mainly limited by the collision effect.

3.2 Efficiency Enhanced MDFH

To further improve spectral efficiency and design flexibility, we can reinforce the transceiver design of the MDFH system by allowing simultaneous transmissions over multiple frequency bands. The modified scheme is referred to as enhanced MDFH (E-MDFH).

3.2.1 The Modified Hopping Frequency Selection Process

Recall that $N_c = 2^{B_c}$ is the number of all available carriers. We split the N_c carriers into N_g non-overlapping groups $\{C_l\}_{l=1}^{N_g}$, with $N_g = 2^{B_g}$, and then each group has $N_f \triangleq N_c/N_g = 2^{B_c - B_g}$ carriers. Specifically, $C_1 = \{f_1, \cdots, f_{N_f}\}$, $C_2 = \{f_{N_f+1}, \cdots, f_{2N_f}\}, \cdots, C_{N_g} = \{f_{N_c - N_f + 1}, \cdots, f_{N_c}\}$. Now we consider to modify the transmitter design in MDFH, such that simultaneous transmissions over multiple frequency bands can be achieved at each hop.

An intuitive method is to employ an independent MDFH scheme within each C_l for $l = 1, \cdots, N_g$. In this case, the frequency hopping processing of each carrier is limited to N_f ($\ll N_c$) carriers.

To maximize the randomness, here we present an alternative approach. We divide the incoming data stream into blocks of length $[N_h(B_c - B_g) + B_s]N_g$. Denote the nth block by X_n, which is further divided into N_g vectors $X_n = [Z_{n,1}, \cdots, Z_{n,N_g}]$. For $m = 1, \cdots, N_g$, each $Z_{n,m}$ contains $N_h(B_c - B_g) + B_s$ bits and can be written as $Z_{n,m} = [D_{n,m}^1, \cdots, D_{n,m}^{N_h}, Y_{n,m}]$. Here each $D_{n,m}^i$ is a bit vector consisting of $(B_c - B_g)$ carrier bits, and the bit vector $Y_{n,m}$ consists of B_s ordinary bits. We adopt the notation "bin2dec" (used in Matlab) to denote the operation of converting a binary vector to a decimal number and "dec2bin" the reverse operation. For $m = 1, \cdots, N_g$, $i = 1, \cdots, N_h$, we define $d_{n,m}^i \triangleq \text{bin2dec}(D_{n,m}^i) + 1$.

Recall that there are N_h hops in one symbol period. At each hop, the signal will be transmitted over N_g carriers simultaneously. For $i = 1, \cdots, N_h$, the frequency index for the mth carrier at the ith hop is defined as:

$$\begin{aligned} I_{n,1}^i &= d_{n,1}^i, && \text{when } m = 1 \\ I_{n,m}^i &= I_{n,m-1}^i + d_{n,m}^i, && \text{when } m = 2, \cdots, N_g. \end{aligned} \tag{3.8}$$

This carrier selection procedure is designed to ensure that (i) all the available carriers are involved in the hopping selection process and (ii) the hopping frequencies have no collisions with each other at any given moment. In fact, at each hop:

$$I_{n,1}^i < I_{n,2}^i < \cdots < I_{n,N_g}^i, \tag{3.9}$$

since $I_{n,1}^i \in [1, N_f]$, $I_{n,2}^i \in [I_{n,1}^i + 1, 2N_f], \cdots, I_{n,N_g}^i \in [I_{n,N_g-1}^i + 1, N_c]$. After the hopping frequencies are determined for all the N_g carriers, each $Y_{n,m}$ ($m = 1, \cdots, N_g$) is mapped to a symbol in constellation Ω and then transmitted through the mth carrier, which hops through frequencies $f_{I_{n,m}^1}, \cdots, f_{I_{n,m}^{N_h}}$. As a result, X_n is now transmitted in one symbol period through simultaneous multi-carrier transmission under the MDFH framework.

3.2.2 Signal Detection

As in MDFH, the receiver in E-MDFH also consists of a bank of N_c bandpass filters. We illustrate the modified signal detection procedure through the extraction of X_n. At the ith hop, instead of searching for the bandpass filter which captures the strongest output in MDFH, we now identify N_g filters which deliver the largest N_g outputs. According to Eq. (3.9), the indices of these N_g bandpass filters are sorted in ascending order, to obtain the estimated indices for $I_{n,m}^i$, that is, $\hat{I}_{n,1}^i < \hat{I}_{n,2}^i < \cdots < \hat{I}_{n,N_g}^i$. Now the carrier-bit vectors can be estimated as:

$$\begin{aligned} \hat{D}_{n,1}^i &= \text{dec2bin}(\hat{I}_{n,1}^i - 1), && \text{when } m = 1 \\ \hat{D}_{n,m}^i &= \text{dec2bin}(\hat{I}_{n,m}^i - \hat{I}_{n,m-1}^i - 1), && \text{when } m = 2, \cdots, N_g. \end{aligned} \tag{3.10}$$

At the same time, each ordinary bit vector $Y_{n,m}$ is estimated from the received signal corresponding to the mth carrier based on bit-wise majority voting, similar to that in MDFH.

Remark 3.1 It is interesting to note that the modified design structure includes both MDFH and OFDM as special cases. In fact, if $N_g = 1$, then E-MDFH is reduced to MDFH. Likewise, if $N_g = N_c$, then E-MDFH can readily be implemented through an OFDM system (see [17]). The advantage of E-MDFH is twofold. First, E-MDFH improves the design flexibility since the transmission scheme can be easily adjusted by tuning the values of B_g or N_g. Second, as a multi-carrier system, E-MDFH can achieve much higher spectral efficiency than MDFH.

3.2.3 Collision-Free MDFH in Multiple Access Environment

E-MDFH can readily be extended to a **collision-free MDFH** scheme to accommodate more users in the multiple access environment, denoted as CF-MDFH. Suppose there are N_u users in the system, all have the same data rate. Note that the block length in the E-MDFH system is L; without loss of generality, we assume that the number of bits assigned to each user is an integer $N_b = \frac{L}{N_u}$. (Otherwise, users may have unequal number of bit assignments so that the total length of the block is L.) Bit streams from different users is mixed through a user multiplexer with an interleaver, as shown in Fig. 3.4a. A secure scrambler is added after the user multiplexer to

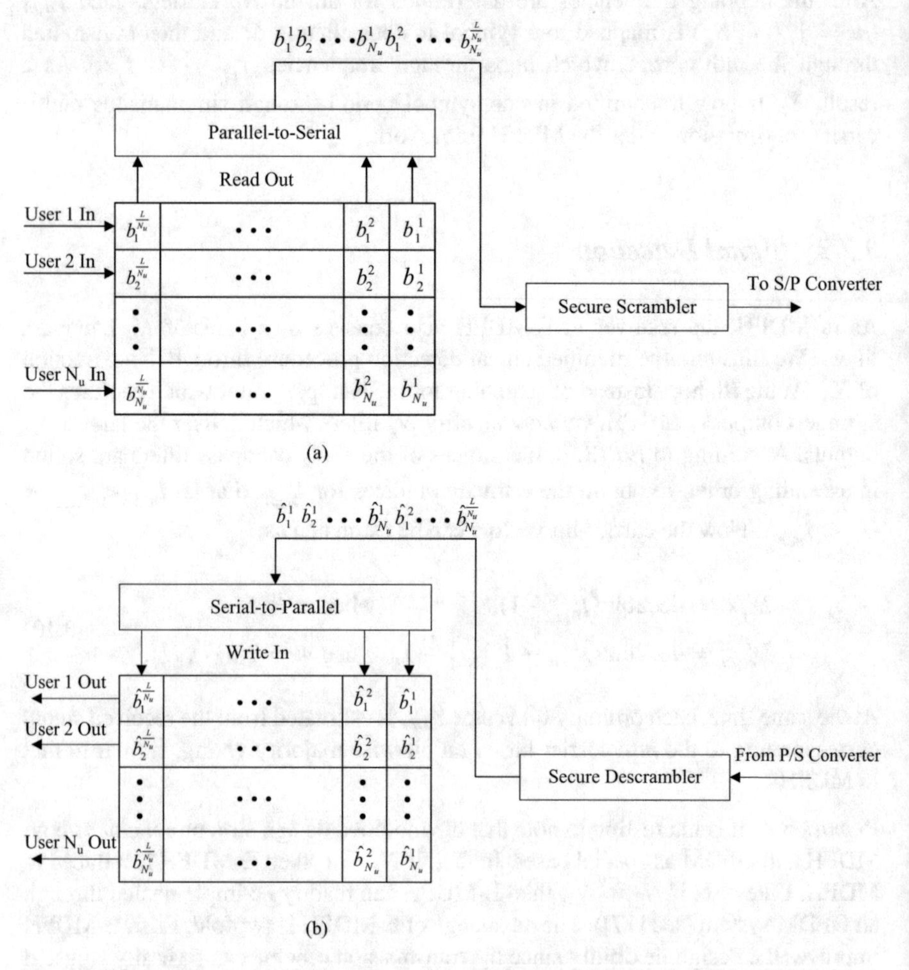

Fig. 3.4 Block-wise user multiplexer and de-multiplexer, designed to process a data block consisted of bits from N_u users. Here b_j^i denotes the ith bit of user j in the block. (**a**) Block-wise user multiplexer. (**b**) Block-wise user de-multiplexer. (© [2009] IEEE. Reprinted, with permission, from Ref. [15])

further randomize the carrier frequencies occupied by each user. A good example of secure scrambler design can be found in [17], where the secure scrambling sequence is obtained by encrypting a PN sequence with the Advanced Encryption Standard (AES). At the receiver, the corresponding de-multiplexer, as shown in Fig. 3.4b, is applied to recover the input bits.

In the CF-MDFH scheme, collision-free multiple access is achieved by sharing the bandwidth resource in frequency domain. CF-MDFH enjoys good *scalability* as the maximum user transmission rate is adaptive, and inversely proportional to the number of users in the system. The carrier frequencies at each hop are jointly determined by the information bits from all the users, and *source anonymity* is ensured by both the interleaver and the secure scrambler. Without the knowledge of the secure scrambler [17], even if the malicious user has a powerful set of equipment that can monitor the whole spectrum and can perfectly recover the carrier bits and ordinary bits, extraction and composition of the desired user's information from the securely scrambled sequence is a forbidden task.

3.3 Performance Analysis for E-MDFH

In this section, we analyze the bit error rate (BER), spectral efficiency of E-MDFH, and also evaluate its performance under hostile jamming.

3.3.1 BER Analysis

Note that in MDFH and E-MDFH, the input bit stream is grouped into carrier bits and ordinary bits, where carrier bits are embedded in hopping frequency selection and the ordinary bits are mapped to symbols in a certain constellation and then transmitted through selected carriers. We will show that *nonuniformity* exists between the carrier bits and the ordinary bits, that is, they have different BER performances.

3.3.1.1 BER of the Carrier Bits

Based on the receiver design in E-MDFH, BER analysis of the carrier bits is analogous to that of noncoherent FSK demodulation. For noncoherent detection of M_F-ary FSK signals, *the probability of symbol error* is given by [18, eqn. (5-4-46), page 310]:

$$P_{s,\mathrm{FSK}}\left(\frac{E_b}{N_0}\right) = \sum_{m=1}^{M_F-1} \binom{M_F-1}{m} \frac{(-1)^{m+1}}{m+1} e^{-\frac{m \log_2 M_F}{(m+1)} \frac{E_b}{N_0}}, \tag{3.11}$$

where $\frac{E_b}{N_0}$ is the bit-level SNR. Let $k_F = \log_2 M_F$, and then *the probability of bit error*, $P_{e,\text{FSK}}$, can be written as:

$$P_{e,\text{FSK}}\left(\frac{E_b}{N_0}\right) = \frac{2^{(k_F-1)}}{2^{k_F}-1} P_{s,\text{FSK}}\left(\frac{E_b}{N_0}\right). \tag{3.12}$$

For an E-MDFH system with N_c channels, $M_F = N_c$, and $k_F = B_c$. Let $\frac{E_b^{(c)}}{N_0}$ and $\frac{E_b^{(o)}}{N_0}$ denote the effective bit-level SNR corresponding to the carrier bits and the ordinary bits, respectively, and $\frac{E_b}{N_0}$ the average bit-level SNR. Recall that in the E-MDFH scheme, the length of each block is $L = [N_h(B_c - B_g) + B_s]2^{B_g}$, out of which there are $B_s 2^{B_g}$ ordinary bits and $N_h(B_c - B_g)2^{B_g}$ carrier bits. Since the carrier bits are embedded in the carrier selection process and do not consume additional transmit power, the average bit-level SNR $\frac{E_b}{N_0}$ is:

$$\frac{E_b}{N_0} = \frac{N_h N_g \bar{E}_s}{N_0[N_h(B_c - B_g) + B_s]2^{B_g}}, \tag{3.13}$$

here $\bar{E}_s$ is the average symbol power. In fact, let $E_1, \cdots, E_{N_t}$ be all the possible power levels in constellation Ω and p_i the probability that the power level of an arbitrary information symbol is E_i, and then the average symbol power $\bar{E}_s$ is given by:

$$\bar{E}_s = \sum_{i=1}^{N_t} p_i E_i, \text{ where } \sum_{i=1}^{N_t} p_i = 1. \tag{3.14}$$

Since each frequency is uniquely identified by B_c bits, and each symbol represents B_s bits, the effective bit-level SNR corresponding to the carrier bits and the ordinary bits can be calculated as:

$$\frac{E_b^{(c)}}{N_0} = \frac{\bar{E}_s}{N_0 B_c}, \quad \frac{E_b^{(o)}}{N_0} = \frac{\bar{E}_s}{N_0 B_s}, \tag{3.15}$$

respectively. Substituting (3.13) into (3.15), it yields that:

$$\frac{E_b^{(c)}}{N_0} = \frac{[N_h(B_c - B_g) + B_s]}{N_h B_c}\frac{E_b}{N_0}, \quad \frac{E_b^{(o)}}{N_0} = \frac{[N_h(B_c - B_g) + B_s]}{N_h B_s}\frac{E_b}{N_0}. \tag{3.16}$$

When $N_g = 1$, E-MDFH is reduced to MDFH. Following (3.11), the BER for the carrier bits in MDFH can be obtained as:

$$P_{e,\text{MDFH}}^{(c)}\left(\frac{E_b}{N_0}\right) = \frac{2^{(B_c-1)}}{2^{B_c}-1}$$
$$\sum_{i=1}^{N_t} p_i \sum_{m=1}^{N_c-1} \binom{N_c-1}{m} \frac{(-1)^{m+1}}{m+1} e^{-\frac{mB_c}{(m+1)}\frac{E_i}{E_s}\frac{E_b^{(c)}}{N_0}}. \tag{3.17}$$

Let $P_{s,\text{MDFH}}^{(c)}$ denote the probability of carrier frequency detection error (corresponding to the symbol error in FSK) in MDFH, and then we have:

$$P_{s,\text{MDFH}}^{(c)}\left(\frac{E_b}{N_0}\right) = \frac{2^{B_c}-1}{2^{(B_c-1)}} P_{e,\text{MDFH}}^{(c)}\left(\frac{E_b}{N_0}\right). \tag{3.18}$$

In the more general case when $N_g > 1$, detection of the carrier bits in E-MDFH is similar to that of differential encoding. Estimation error in one carrier index may cause detection errors in two neighboring carrier bit blocks. Denote the probability of carrier frequency detection error in E-MDFH as $P_{\text{E-MDFH}}^{(c)}$; it follows from (3.11) that:

$$P_{\text{E-MDFH}}^{(c)}\left(\frac{E_b}{N_0}\right) = \sum_{i=1}^{N_t} p_i \sum_{m=1}^{N_c-1} \binom{N_c-1}{m} \frac{(-1)^{m+1}}{m+1} e^{-\frac{mB_c}{(m+1)}\frac{E_i}{E_s}\frac{E_b^{(c)}}{N_0}}. \tag{3.19}$$

Please note that $\frac{E_b}{N_0}$ in E-MDFH is different from that in MDFH due to different block structures.

Recall that $i = 1, \cdots, N_h$, $m = 1, \cdots, N_g$, $I_{n,m}^i$ denotes the frequency index for the mth carrier at the ith hop of the nth symbol period. At each hop, $I_{n,m}^i$ should satisfy $I_{n,1}^i < I_{n,2}^i < \cdots < I_{n,N_g}^i$. For signal detection, after each individual carrier index is estimated, they are then sorted in ascending order to recover $I_{n,m}^i$. An error in the carrier index estimation may further introduce errors in the sorting process and hence has negative impact on the index estimation for more than one $I_{n,m}^i$. Therefore, if $P_{\text{E-MDFH}}^{(I)}\left(\frac{E_b}{N_0}\right)$ denotes the average probability that an index $I_{n,m}^i$ is incorrectly estimated, then we have $P_{\text{E-MDFH}}^{(I)}\left(\frac{E_b}{N_0}\right) \geq P_{\text{E-MDFH}}^{(c)}\left(\frac{E_b}{N_0}\right)$.

Since $d_{n,1}^i$ is uniquely determined by $I_{n,1}^i$, it is clear that the probability of error in estimating $d_{n,1}^i$ is $P_{\text{E-MDFH}}^{(I)}\left(\frac{E_b}{N_0}\right)$. Because there is a total of N_g carriers hopping over the entire bandwidth simultaneously, this type of error only counts for $\frac{1}{N_g}$ of all the possible errors. Thus, the BER for the first carrier is $\frac{1}{N_g}\frac{2^{(B_c-B_g-1)}}{2^{(B_c-B_g)}-1} P_{\text{E-MDFH}}^{(I)}\left(\frac{E_b}{N_0}\right)$. For the rest $\{d_{n,m}^i\}_{m=2}^{N_g}$, each $d_{n,m}^i$ relies on both $I_{n,m}^i$ and $I_{n,m-1}^i$, according to the generation rule for the frequency indices in Eq. (3.8). Therefore, the probability that $d_{n,m}^i$ is correctly detected, for $m = 2, \cdots, N_g$, is $\left[1 - P_{\text{E-MDFH}}^{(I)}\left(\frac{E_b}{N_0}\right)\right]^2$. In other words, the probability that $d_{n,m}^i$ is not correctly detected is $1-\left[1-P^{(I)}_{\text{E-MDFH}}\left(\frac{E_b}{N_0}\right)\right]^2$. Considering that the weight for this type of error is $\frac{N_g-1}{N_g}$, we have the overall BER for the carrier bits as the weighted summation of two types of errors:

$$P^{(c)}_{e,\text{E-MDFH}}\left(\frac{E_b}{N_0}\right)=\frac{2^{(B_c-B_g-1)}}{2^{(B_c-B_g)}-1}\left\{\frac{1}{N_g}P^{(I)}_{\text{E-MDFH}}\left(\frac{E_b}{N_0}\right)\right. \\ \left.+\frac{N_g-1}{N_g}\left\{1-\left[1-P^{(I)}_{\text{E-MDFH}}\left(\frac{E_b}{N_0}\right)\right]^2\right\}\right\}.$$

Note that $P^{(I)}_{\text{E-MDFH}}\left(\frac{E_b}{N_0}\right)\geq P^{(c)}_{\text{E-MDFH}}\left(\frac{E_b}{N_0}\right)$, a lower bound of the BER for the carrier bits can thus be obtained as:

$$P^{(c),L}_{e,\text{E-MDFH}}\left(\frac{E_b}{N_0}\right)=\frac{2^{(B_c-B_g-1)}}{2^{(B_c-B_g)}-1}\left\{\frac{1}{N_g}P^{(c)}_{\text{E-MDFH}}\left(\frac{E_b}{N_0}\right)\right. \\ \left.+\frac{N_g-1}{N_g}\left\{1-\left[1-P^{(c)}_{\text{E-MDFH}}\left(\frac{E_b}{N_0}\right)\right]^2\right\}\right\}. \tag{3.20}$$

Furthermore, the probability that all the indices $\{I^i_{n,m}\}^{N_g}_{m=1}$ are correctly estimated is $\left[1-P^{(c)}_{\text{E-MDFH}}\left(\frac{E_b}{N_0}\right)\right]^{N_g}$. In this case, all the carrier bits can be perfectly recovered. If we assume that any estimation error in $\{I^i_{n,m}\}^{N_g}_{m=1}$ will incur detection errors in all the carrier bit blocks, then an upper bound of the BER for the carrier bits can be obtained as:

$$P^{(c),U}_{e,\text{E-MDFH}}\left(\frac{E_b}{N_0}\right)=\frac{2^{(B_c-B_g-1)}}{2^{(B_c-B_g)}-1}\left\{1-\left[1-P^{(c)}_{\text{E-MDFH}}\left(\frac{E_b}{N_0}\right)\right]^{N_g}\right\}. \tag{3.21}$$

To summarize our discussions above, we have:

Proposition 3.1 *In E-MDFH, the BER for the carrier bits, $P^{(c)}_{e,\text{E-MDFH}}$, is bounded by:*

$$P^{(c),L}_{e,\text{E-MDFH}}\left(\frac{E_b}{N_0}\right)\leq P^{(c)}_{e,\text{E-MDFH}}\left(\frac{E_b}{N_0}\right)\leq P^{(c),U}_{e,\text{E-MDFH}}\left(\frac{E_b}{N_0}\right). \tag{3.22}$$

Accordingly, the probability of error on the estimation of $d^i_{n,m}$, $P^{(d)}_{\text{E-MDFH}}$, for $i=1,\cdots,N_h$, $m=1,\cdots,N_g$ is bounded by:

$$\underbrace{\frac{2^{(B_c-B_g)}-1}{2^{(B_c-B_g-1)}} P_{e,\text{E-MDFH}}^{(c),L}\left(\frac{E_b}{N_0}\right)}_{\triangleq P_{s,\text{E-MDFH}}^{(c),L}\left(\frac{E_b}{N_0}\right)} \le P_{\text{E-MDFH}}^{(d)}\left(\frac{E_b}{N_0}\right)$$

$$\le \underbrace{\frac{2^{(B_c-B_g)}-1}{2^{(B_c-B_g-1)}} P_{e,\text{E-MDFH}}^{(c),U}\left(\frac{E_b}{N_0}\right)}_{\triangleq P_{s,\text{E-MDFH}}^{(c),U}\left(\frac{E_b}{N_0}\right)}.$$

3.3.1.2 BER of the Ordinary Bits

BER of the ordinary bits is determined by the modulation scheme used in the system. If FSK is utilized, then the BER of the ordinary bits can be calculated in a similar manner as that of the carrier bits. In the following, we consider the case of transmitting the ordinary bits through M-ary QAM. We start with MDFH, which is easier to analyze, and then extend the results to E-MDFH.

Recall that if $M = 2^{B_s}$, where B_s is an even integer, the probability of symbol error for M-ary QAM is [18, eqn. (5-2-78) & (5-2-79)]:

$$P_{s,\text{MQAM}}\left(\frac{E_b}{N_0}\right) = 1 - \left[1 - 2\left(1 - \frac{1}{\sqrt{M}}\right) Q\left(\sqrt{\frac{3\log_2 M}{(M-1)}\frac{E_b}{N_0}}\right)\right]^2, \quad (3.23)$$

where $Q(x) = \frac{1}{\sqrt{2\pi}}\int_x^\infty e^{-\frac{t^2}{2}}dt$. Taking 16-QAM as an example, we have $P_{s,16\text{-QAM}}\left(\frac{E_b}{N_0}\right) = \frac{9}{4}\left[\frac{4}{3} - Q\left(\sqrt{\frac{4}{5}\frac{E_b}{N_0}}\right)\right] Q\left(\sqrt{\frac{4}{5}\frac{E_b}{N_0}}\right)$.

In MDFH, each QAM symbol undergoes N_h hops, and we assume that N_h is odd. For signal detection, we first estimate the QAM symbol independently for each hop, and then apply *bit-wise majority voting* for the N_h estimates to make the final decision. Accordingly, the BER of the ordinary bits, $P_{e,\text{MDFH}}^{(o)}$, can be calculated as follows:

(i) *BER analysis at each individual hop.* At each hop, the bit error can be classified into two types.

- *Type I error: bit is in error given that the carrier frequency is correctly detected.* When the carrier frequency is detected correctly, for which the probability is $\left(1 - P_{s,\text{MDFH}}^{(c)}\left(\frac{E_b}{N_0}\right)\right)$, the probability of bit error can be calculated based on the BER of coherently detected M-ary QAM, given by $P_{e1} \triangleq P_{e,\text{MQAM}}\left(\frac{E_b^{(o)}}{N_0}\right)$. Here, $\frac{E_b}{N_0} = \frac{N_h B_s}{N_h B_c + B_s}\frac{E_b^{(o)}}{N_0}$.

- *Type II error: bit is in error when the carrier frequency is not correctly detected.* When the carrier frequency is not correctly detected, for which the probability is $P^{(c)}_{s,\text{MDFH}}\left(\frac{E_b}{N_0}\right)$, it is reasonable to assume that the probability of bit error is $P_{e2} \triangleq \frac{1}{2}$.

(ii) *Average BER calculation based on majority voting.* In MDFH, each QAM symbol is transmitted through N_h hops. As a result, an error in a particular bit location is caused by at least $\lceil \frac{N_h}{2} \rceil$ unsuccessful recovery, where $\lceil x \rceil$ denotes the smallest integer greater than or equal to x. Let $P_{e,i}, i = 0, 1, \cdots, N_h$, be the conditional probability of bit error given that i out of N_h carrier frequencies are not correctly detected (i.e., $N_h - i$ carrier frequencies are correctly detected). Let j denote the number of unsuccessful bit recovery:

$$\begin{aligned}
&P_{e,i}\left(\frac{E_b}{N_0}\right)\\
&= \sum_{j=\lceil \frac{N_h}{2} \rceil}^{N_h} \text{Prob}\{\text{errors in } j \text{ out of } N_h \text{ hops} \mid i\\
&\quad \text{carriers are not correctly detected}\}\\
&= \sum_{j=\lceil \frac{N_h}{2} \rceil}^{N_h} \sum_{k=0}^{j} \text{Prob}\{k \text{ type I errors and } j-k \text{ type II errors} \mid i\\
&\quad \text{carriers are not correctly detected}\}\\
&= \sum_{j=\lceil \frac{N_h}{2} \rceil}^{N_h} \sum_{k=0}^{j} \binom{N_h - i}{k} (P_{e1})^k (1 - P_{e1})^{N_h-i-k} \binom{i}{j-k} (P_{e2})^{j-k}\\
&\quad (1 - P_{e2})^{i-j+k}.
\end{aligned} \tag{3.24}$$

Here we adopt the convention $\binom{n}{m} = 0$ when $n < m$.

Taking the effect of the majority voting into consideration, the error probability for the ordinary bits, $P^{(o)}_{e,\text{MDFH}}$, is given by:

$$\begin{aligned}
&P^{(o)}_{e,\text{MDFH}}\left(\frac{E_b}{N_0}\right)\\
&= \sum_{i=0}^{N_h} \text{Prob}\{\text{the bit is in error} \mid i \text{ carriers are wrong }\}\text{Prob}\{i \text{ carriers are wrong }\}\\
&= \sum_{i=0}^{N_h} \binom{N_h}{i} \left[P^{(c)}_{s,\text{MDFH}}\left(\frac{E_b}{N_0}\right)\right]^i \left[1 - P^{(c)}_{s,\text{MDFH}}\left(\frac{E_b}{N_0}\right)\right]^{N_h-i} P_{e,i}\left(\frac{E_b}{N_0}\right).
\end{aligned} \tag{3.25}$$

To determine the bit error probability for the ordinary bits in E-MDFH, we proceed in a similar manner as in MDFH, except for the use of different error probabilities in the detection of carrier frequency. More specifically, lower and upper bounds of the probability of bit error for the ordinary bits can be obtained by substituting $P_{s,\text{E-MDFH}}^{(c),L}\left(\frac{E_b}{N_0}\right)$ and $P_{s,\text{E-MDFH}}^{(c),U}\left(\frac{E_b}{N_0}\right)$ for $P_{s,\text{MDFH}}^{(c)}\left(\frac{E_b}{N_0}\right)$ in (3.25), respectively.

3.3.1.3 Overall BER for E-MDFH

The overall BER of the E-MDFH scheme is calculated as the linear combination of $P_{e,\text{E-MDFH}}^{(c)}$ and $P_{e,\text{E-MDFH}}^{(o)}$ based on the number of carrier bits and the number of ordinary bits in each block,

$$P_{e,\text{E-MDFH}}\left(\frac{E_b}{N_0}\right) = \frac{N_h(B_c - B_g)}{N_h(B_c - B_g) + B_s} P_{e,\text{E-MDFH}}^{(c)}\left(\frac{E_b}{N_0}\right) + \frac{B_s}{N_h(B_c - B_g) + B_s} P_{e,\text{E-MDFH}}^{(o)}\left(\frac{E_b}{N_0}\right). \tag{3.26}$$

3.3.2 *Spectral Efficiency Analysis of MDFH*

In this section, we compare the spectral efficiency of the E-MDFH scheme with that of the conventional FH scheme.

We start with the *single user case*, that is, there is only one user in both systems and no collisions need to be taken into consideration. Recall that T_s denotes the symbol period and T_h the hopping duration, and $N_h = T_s/T_h$ ($N_h \geq 1$) is the number of hops per symbol period. For fair comparison, we assume that both systems have the same symbol period T_s, the same number of hops per symbol, N_h, and use constellation(s) of the same size M, i.e., the number of bits per symbol is $B_s = \log_2 M$. Let $R_s \triangleq 1/T_s$ be the symbol rate. Accordingly, the bit rate of conventional FH can be expressed as:

$$R_{b,FH} = B_s R_s \text{ bits/second.} \tag{3.27}$$

Recall that the data rate of E-MDFH $R_{b,\text{E-MDFH}}$ is $[N_h(B_c - B_g) + B_s]2^{B_g}$ bits every symbol period. That is:

$$R_{b,\text{E-MDFH}} = [N_h(B_c - B_g) + B_s]2^{B_g} R_s \text{ bits/second.} \tag{3.28}$$

Given N_h, B_c, B_s, an interesting question is to find the optimal B_g that maximizes $R_{b,\text{E-MDFH}}$. By solving $\frac{dR_{b,\text{E-MDFH}}}{dB_g} = 0$, we have $B_g = B_c + \frac{B_s}{N_h} - \frac{1}{\ln 2}$. Note that B_g must be an integer and $B_g \in [0, B_c]$, we have the following results:

Proposition 3.2 *Let* $B_g^{\perp} \triangleq \max\{0, \lfloor B_c + \frac{B_s}{N_h} - \frac{1}{\ln 2} \rfloor\}$ *and* $B_g^{\top} \triangleq \min\{B_c, \lceil B_c + \frac{B_s}{N_h} - \frac{1}{\ln 2} \rceil\}$*, where* $\lfloor x \rfloor$ *denotes the largest integer less than or equal to x and* $\lceil x \rceil$ *the smallest integer greater than or equal to x. The optimal value of* B_g*, denoted by* B_g^**, that maximizes the throughput of the E-MDFH is given by:*

$$B_g^* = \begin{cases} B_g^{\perp}, \text{ if } \frac{[N_h(B_c - B_g^{\perp}) + B_s]2^{B_g^{\perp}}}{[N_h(B_c - B_g^{\top}) + B_s]2^{B_g^{\top}}} > 1, \\ B_g^{\top}, \text{ otherwise.} \end{cases} \tag{3.29}$$

Given that the total bandwidth $W_B = c_0 \frac{N_c}{T_h}$, where c_0 is a constant, the *spectral efficiency* (in bits/second/Hz) of the conventional fast FH and E-MDFH are given by

$$\eta_{FH} = \frac{R_{b,FH}}{W_B} = \frac{B_s}{c_0 N_c N_h}, \tag{3.30}$$

$$\eta_{\text{E-MDFH}} = \frac{R_{b,\text{E-MDFH}}}{W_B} = \frac{[N_h(B_c - B_g^*) + B_s]2^{B_g^*}}{c_0 N_c N_h}. \tag{3.31}$$

It is obvious that we always have $\eta_{\text{E-MDFH}} > \eta_{FH}$. That is, E-MDFH is always much more efficient than the conventional fast FH scheme.

Next, we explore the more general case where there are *multiple users* in both systems.

Consider a conventional fast FH system with $N_u (>1)$ users, each transmitting 2^{B_c}-ary MFSK signals over N_c frequencies. At the receiver, after de-hopping, each user creates a $2^{B_c} \times N_h$ decision table. For $i = 1, \cdots, 2^{B_c}$ and $j = 1, \cdots, N_h$, if the receiver decides a transmission is present on the ith carrier frequency during the jth hop duration, that is, if the output of the corresponding envelope detector exceeds some threshold, then the entry in the ith row and jth column of the receiver matrix is filled; otherwise, it is left as blank [2, 19].

Note that in the conventional FHMA systems, different users hop in an independent manner, any one of the $2^{B_c} - 1$ carrier frequencies not occupied by the desired user may be filled incorrectly as a result of a tone transmitted by an interfering user. This event, known as *insertion*, occurs with probability $p = 1 - \left(1 - \frac{1}{2^{B_c}}\right)^{N_u - 1}$. Moreover, additive noise may also result in insertions to the receiver matrix. As a result, the overall insertion probability P_I is given by [2]

$$P_I = p + (1 - p)e^{-\theta^2 \frac{B_c}{N_h} \frac{E_b}{N_0}}, \tag{3.32}$$

where θ is the normalized threshold.

On the other hand, because of channel distortion and noise effect, the decision statistic may fall below the threshold, a tone from the desired user may not appear among the receiver's positive decisions, resulting in a *deletion* in the decision matrix. The probability that a *deletion* occurs, denoted by P_D, is given by [2]:

$$P_D = \frac{p}{3} + (1 - \frac{p}{3})\left[1 - Q\left(\sqrt{\frac{2B_c}{N_h}\frac{E_b}{N_0}}, \theta\sqrt{\frac{2B_c}{N_h}\frac{E_b}{N_0}}\right)\right], \tag{3.33}$$

where $Q(x, y)$ is the Marcum Q-function [20].

In the ideal case, the row of the receive matrix, corresponding to the symbol transmitted by the desired user, will be fully filled. If it is the only full row, then the decoding is perfect. However, due to the presence of multiple access interference and additive white noise, insertion as well as deletion in the receiver matrix may occur. When the filled entries in another row is more than that of the desired row, a symbol error occurs. When two or more rows in the receiver matrix have the same number of entries, random decision (based on the toss of a coin) has to be used for MFSK symbol detection.

Let $P_d(i)$ denote the probability that i entries are filled on the desired row of the receiver matrix and $P_o(j)$ the probability that j entries are filled on an undesired row, and then we have: $P_d(i) = \binom{N_c}{i}(1 - P_D)^i P_D^{N_h - i}$, and $P_o(j) = \binom{N_c}{j}(1 - P_I)^{N_h - j} P_I^j$.

Let $P_{e,\text{FHMA}}^{(u)}$ denote the bit-error-rate of the desired user in the conventional FHMA system. Let N_w be the total number of undesired rows, and then $N_w = 2^{B_c} - 1$. Let j_n be the number of filled entries in an undesired row f_n and define $j_0 \triangleq \max\{j_n\}$, where the maximum operation is taken over all the undesired rows. As can be seen, a detection error occurs whenever $j_0 \geq i$. Since the probability that $j_n = j_0$ occurs in two or more undesired rows is very low, here we only consider the case where only one undesired row has j_0 entries. As a result, $P_{e,\text{FHMA}}^{(u)}$ can be approximated as:

$$P_{e,\text{FHMA}}^{(u)} \approx \frac{2^{(B_c-1)}}{2^{B_c} - 1}\binom{N_w}{1}\left(\text{Prob}\,\{j > i\} + \frac{1}{2}\text{Prob}\{\text{j} = \text{i}\}\right), \tag{3.34}$$

$$= 2^{(B_c-1)}\left[\sum_{j=1}^{N_h} P_o(j)\sum_{i=0}^{j-1} P_d(i) + \frac{1}{2}\sum_{j=1}^{N_h} P_o(j)P_d(j)\right], \tag{3.35}$$

where the factor of $\frac{1}{2}$ is the probability that the fair coin toss favors the wrong decision rather than the correct one.

For the E-MDFH scheme, there is no multiple access interference; therefore, the average BER in (3.26) is directly applicable to the multiuser case. From the spectral efficiency perspective, we need to compare the total information bits allowed to be

Table 3.1 Comparison of information bit rate between the conventional fast FH system and the E-MDFH scheme for various hop rates. $N_c = 64$, $B_s = 4$, $B_g = 2$. (© [2009] IEEE. Reprinted, with permission, from Ref. [15])

	Case 1	Case 2	Case 3	Case 4	Case 5	Case 6
N_h	3	3	5	5	7	7
Required BER	10^{-3}	10^{-4}	10^{-4}	10^{-5}	10^{-4}	10^{-5}
Conventional fast FH (bits/T_s)	30	18	30	24	42	30
E-MDFH (bits/T_s)	64	64	96	96	128	128

transmitted under the same BER and bandwidth requirements (i.e., the same hop rate). We illustrate the system performance through a numerical example below. It is shown that E-MDFH can increase the spectral efficiency by multiple times.

Example (Spectral Efficiency of E-MDFH under AWGN) Assume $N_c = 64$ (i.e., $B_c = 6$), $B_s = 4$, and $B_g = 2$. The overall result is listed in Table 3.1. As can be seen, E-MDFH can achieve a spectral efficiency up to four times higher than that of the conventional FHMA system.

3.3.3 *Performance Analysis of MDFH Under Hostile Jamming*

In this section, we evaluate the performance of MDFH under both noise jamming and disguised jamming. Recall that *disguised jamming* denotes the case where the jamming is highly correlated with the signal and has a power level close to the signal power. More specifically, let $s(t)$ and $J(t)$ denote user's signal and jamming interference, respectively. Define

$$\rho = \frac{1}{T_0\sqrt{P_s P_J}} \int_{t_1}^{t_2} s(t)J^*(t)dt \tag{3.36}$$

as the normalized cross-correlation coefficient of $s(t)$ and $J(t)$ over the time period $[t_1, t_2]$, where $T_0 = t_2 - t_1$, $P_s = \frac{1}{T_0}\int_{t_1}^{t_2} |s(t)|^2 dt$ and $P_J = \frac{1}{T_0}\int_{t_1}^{t_2} |J(t)|^2 dt$. We say that $J(t)$ is a disguised jamming to signal $s(t)$ over $[t_1, t_2]$ if

1. $J(t)$ and $s(t)$ are highly correlated. More specifically, $|\rho| > \rho_0$, where ρ_0 is an application-oriented predefined correlation threshold.
2. The jamming-to-signal ratio (JSR) is close to 0 dB. More specifically, $|\frac{P_J}{P_s} - 1| < \epsilon_P$, where ϵ_P is an application-oriented predefined jamming-to-signal ratio threshold.

In this section, we consider disguised jamming over each hopping period, that is, $[t_1, t_2] = [mT_h, (m+1)T_h]$ for some integer m. In *the worst case*, the constellation Ω and the pulse shaping filter of the information signal are known

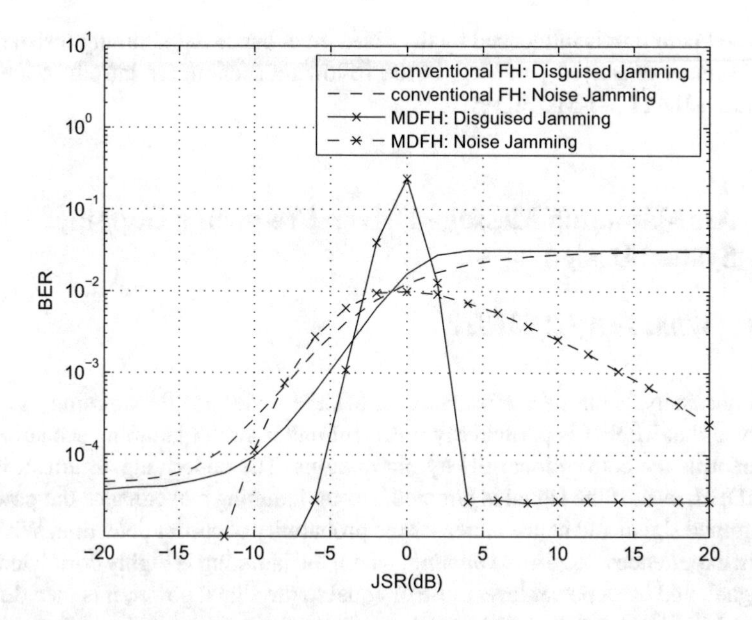

Fig. 3.5 Performance comparison under single band jamming, $E_b/N_0 = 10\,\text{dB}$, $N_c = 64$, $N_h = 3$. MDFH uses QPSK modulation and conventional FH uses 4-FSK modulation. In this case, the spectral efficiency of MDFH is roughly 3.3 times that of conventional FH. (© [2013] IEEE. Reprinted, with permission, from Ref. [21])

to the jammer; the jammer can then disguise itself by transmitting symbols from Ω over a fake channel using the same power level. That is, $J(t) = e^{i\theta}s(t)$ for some phase θ.

We compare the performance of MDFH with that of conventional FH in AWGN channels, under both noise jamming and disguised jamming. The result (with no channel coding) is shown in Fig. 3.5. The jamming-to-signal ratio is defined as $JSR = \frac{P_J}{P_s}$, where P_J and P_s denote the jamming power and signal power per hop, respectively. As can be seen, *MDFH delivers excellent performance under strong jamming scenarios (i.e., $JSR \gg 1$)* and outperforms conventional FH by big margins. Note that in this case, the spectral efficiency of MDFH is 3.3 times that of conventional FH. The underlying argument is that when the jamming power is much stronger than the signal power, jamming can be easily distinguished from the true signal when they are in different bands; even if jamming collides with the signal, the true carrier frequency can still be detected as jamming can even enhance the power of the jammed channel and hence increases the probability of carrier detection. For conventional FH, on the other hand, once the jamming power reaches a certain level, the system performance is mainly limited by the probability that the signal is jammed.

However, we also notice that under disguised jamming, the system experiences considerable performance losses, since it is difficult for the MDFH receiver to distinguish disguised jamming from the true signal. The sensitivity of MDFH to disguised jamming is influenced by the SNR. To enhance the jamming resistance of MDFH under disguised jamming, in the following section, we introduce the anti-jamming MDFH (AJ-MDFH) system.

3.4 Anti-jamming Message-Driven Frequency Hopping: System Design

3.4.1 What Is AJ-MDFH?

From our analysis on the performance of MDFH under hostile jamming, it can be observed that MDFH is particularly powerful under strong jamming scenarios and outperforms the conventional FH by big margins. The underlying argument is that for MDFH, even if the signal is jammed, strong jamming can enhance the power of the jammed signal and hence increases the probability of carrier detection. When the system experiences *disguised jamming*, where the jamming is highly correlated with the signal, and has a power level close or equal to the signal power, it is then difficult for the MDFH receiver to distinguish jamming from the true signal, resulting in performance losses. As will be shown in the next section, this is essentially due to the existence of symmetricity between the jamming interference and the authorized signal.

To overcome the drawback of MDFH, in this section, we present an anti-jamming MDFH (AJ-MDFH) scheme [21, 22]. The main idea is to insert some signal identification (ID) information during the transmission process. This secure ID information is generated through the Advanced Encryption Standard (AES) [23] using the shared secret between the transmitter and the receiver. The ID information can be exploited by the receiver to locate the true carrier frequency. Moreover, protected by AES, it is computationally infeasible for malicious users to recover the ID sequence.

The major difference with MDFH is that in AJ-MDFH, we add shared randomness between the transmitter and the receiver to break the symmetricity between the jamming interference and authorized signal. More specifically, in AJ-MDFH, secure ID signals are introduced to distinguish the true information channel from the disguised channels invoked by jamming interference. Our analysis indicates that comparing with MDFH, AJ-MDFH can effectively reduce the performance degradation caused by disguised jamming. At the same time, its spectral efficiency is very close to that of MDFH, which is several times higher than that of conventional FH.

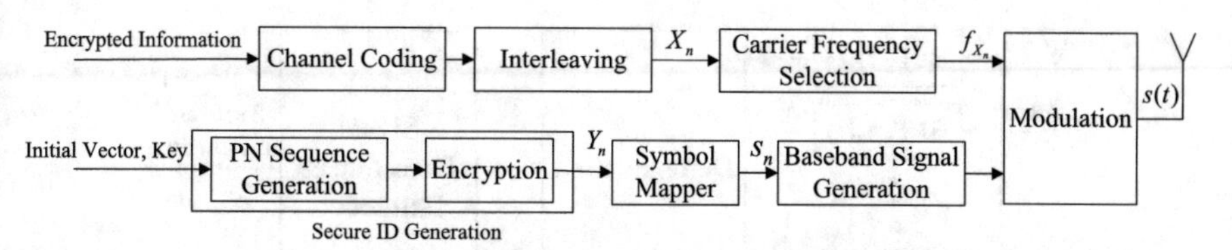

Fig. 3.6 AJ-MDFH transmitter structure. (© [2013] IEEE. Reprinted, with permission, from Ref. [21])

3.4.2 Transmitter Design

The main idea here is to insert some signal identification (ID) information during the transmission process. This secure ID information is generated through a cryptographic algorithm using the shared secret between the transmitter and the receiver and can be used by the receiver to locate the true carrier frequency. Our design goal is to reinforce jamming resistance without sacrificing too much on spectral efficiency.

The transmitter structure of AJ-MDFH is illustrated in Fig. 3.6. Each user is assigned a secure ID sequence. We replace the ordinary bits in MDFH with the ID bits. In order to prevent impersonate attack, each user's ID sequence needs to be kept secret from the malicious jammer. The ID sequence can be generated using two steps as in [17]: (i) generate a pseudo-random binary sequence using a linear feedback shift register (LFSR); and (ii) take the output of LFSR as the plaintext, and feed it into the Advanced Encryption Standard (AES) [23] encrypter. The AES output is then used as our ID sequence.

Recall that $B_c = \log_2 N_c$ and $B_s = \log_2 M$, where N_c is the number of channels and M is the constellation size. We divide the source information into blocks of size B_c and divide the ID sequence into blocks of size B_s. Denote the nth source information block and ID block as X_n and Y_n, respectively. Let f_{X_n} be the carrier frequency corresponding to X_n and s_n the symbol corresponding to ID bit vector Y_n. *It should be noted that the ID symbol is refreshed at each hopping period.* The transmitted signal can then be represented as:

$$\begin{aligned} s(t) &= \sqrt{2}\mathrm{Re}\left\{\sum_{n=-\infty}^{\infty} s_n g(t-nT_h)e^{j2\pi f_{X_n} t}\right\} \\ &= \sqrt{2}\mathrm{Re}\left\{\sum_{n=-\infty}^{\infty}\sum_{i=1}^{N_c} \alpha_{i,n} s_n g(t-nT_h)e^{j2\pi f_i t}\right\}, \end{aligned} \tag{3.37}$$

where T_h is the hop duration and $g(t)$ is the pulse shaping filter:

$$\alpha_{i,n} = \begin{cases} 1 \text{ if } f_{X_n} = f_i, \\ 0 \text{ otherwise.} \end{cases}$$

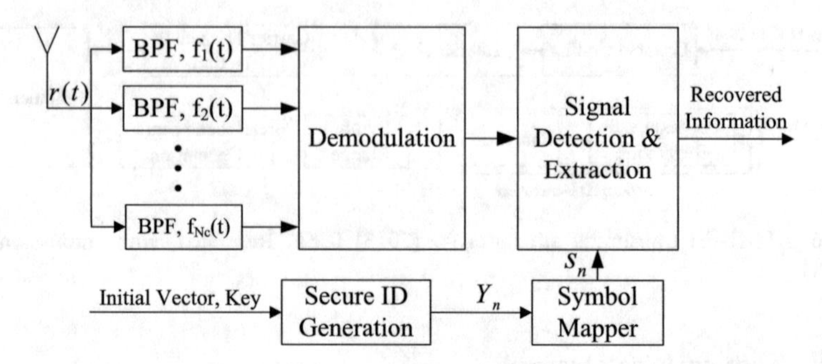

Fig. 3.7 AJ-MDFH receiver structure. (© [2013] IEEE. Reprinted, with permission, from Ref. [21])

3.4.3 Receiver Design

The receiver structure for AJ-MDFH is shown in Fig. 3.7. For each hop, the received signal is first fed into the bandpass filter bank. The output of the filter bank is demodulated and then used for carrier bits (i.e., the information bits) detection.

3.4.3.1 Demodulation

Let $s(t)$, $J(t)$, and $n(t)$ denote the ID signal, the jamming, and the noise, respectively. For AWGN channels, the received signal can be represented as:

$$r(t) = s(t) + J(t) + n(t). \tag{3.38}$$

For $i = 1, 2, \cdots, N_c$, the output of the ith ideal bandpass filter $\mathrm{f}_i(t)$ is $r_i(t) = \mathrm{f}_i(t) * r(t)$. For demodulation, $r_i(t)$ is first shifted back to the baseband and then passed through a matched filter. At the nth hopping period, for $i = 1, \cdots, N_c$, the sampled matched filter output corresponding to channel i can be expressed as:

$$r_{i,n} = \alpha_{i,n} s_n + \beta_{i,n} J_{i,n} + n_{i,n}, \tag{3.39}$$

where s_n, $J_{i,n}$, and $n_{i,n}$ correspond to the ID symbol, the jamming interference, and the noise, respectively; $\alpha_{i,n}, \beta_{i,n} \in \{0, 1\}$ are binary indicators for the presence of ID signal and jamming, respectively. Note that the true information is carried in $\alpha_{i,n}$.

3.4.3.2 Signal Detection and Extraction

Signal detection and extraction is performed for each hopping period. *For notation simplicity, without loss of generality, we omit the subscript n in* (3.39). That is, for a particular hopping period, (3.39) is reduced to:

$$r_i = \alpha_i s + \beta_i J_i + n_i, \quad \text{for } i = 1, \cdots, N_c. \tag{3.40}$$

Define $\mathbf{r} = (r_1, \ldots, r_{N_c})$, $\boldsymbol{\alpha} = (\alpha_1, \ldots, \alpha_{N_c})$, $\boldsymbol{\beta} = (\beta_1, \ldots, \beta_{N_c})$, $\mathbf{J} = (J_1, \ldots, J_{N_c})$ and $\mathbf{n} = (n_1, \ldots, n_{N_c})$, and then (3.40) can be rewritten in vector form as:

$$\mathbf{r} = s\boldsymbol{\alpha} + \boldsymbol{\beta} \cdot \mathbf{J} + \mathbf{n}. \tag{3.41}$$

For single-carrier AJ-MDFH, at each hopping period, one and only one item in $\boldsymbol{\alpha}$ is nonzero. In this case, there are N_c possible information vectors: $\boldsymbol{\alpha}_1 = (1, 0, \ldots, 0)$, $\boldsymbol{\alpha}_2 = (0, 1, \ldots, 0)$, $\cdots$, $\boldsymbol{\alpha}_{N_c} = (0, 0, \ldots, 1)$. If $\boldsymbol{\alpha}_k$ is selected, and the binary expression of k is $b_0 b_1 \cdots b_{B_c - 1}$, with $B_c = \lfloor \log_2 N_c \rfloor$, then the estimated information sequence is $b_0 b_1 \cdots b_{B_c-1}$.

At each hopping period, the information symbol $\boldsymbol{\alpha}$, or equivalently, the hopping frequency index k, needs to be estimated based on the received signal and the secure ID information which can be regenerated at the receiver through the shared secret. When the input information vectors are equiprobable, that is, $P(\boldsymbol{\alpha}_i) = \frac{1}{N_c}$ for $i = 1, 2, \ldots, N_c$, the MAP (maximum a posteriori probability) detector is reduced to the ML (maximum likelihood) detector. For the ML detector, the hopping frequency index $\hat{k}$ can be estimated as:

$$\hat{k} = \arg \max_{1 \le i \le N_c} Pr\{\mathbf{r} | \boldsymbol{\alpha}_i\}. \tag{3.42}$$

When $n_1, \ldots, n_{N_c}$, $J_1, \ldots, J_{N_c}$ are all statistically independent, $r_1, \ldots, r_{N_c}$ are also independent. In this case, the joint ML detector in (3.42) can be decomposed as:

$$\begin{aligned}
\hat{k} &= \arg \max_{1 \le i \le N_c} \prod_{j=1}^{N_c} Pr\{r_j | \boldsymbol{\alpha}_i\} \\
&= \arg \max_{1 \le i \le N_c} \prod_{j=1, j \ne i}^{N_c} Pr\{r_j | \alpha_j = 0\} \cdot Pr\{r_i | \alpha_i = 1\} \\
&= \arg \max_{1 \le i \le N_c} \prod_{j=1}^{N_c} Pr\{r_j | \alpha_j = 0\} \cdot \frac{Pr\{r_i | \alpha_i = 1\}}{Pr\{r_i | \alpha_i = 0\}}.
\end{aligned} \tag{3.43}$$

Since $\prod_{j=1}^{N_c} Pr\{r_j | \alpha_j = 0\}$ is independent of i, (3.43) can be further reduced to the likelihood ratio test:

$$\hat{k} = \arg \max_{1 \le i \le N_c} \frac{Pr\{r_i | \alpha_i = 1\}}{Pr\{r_i | \alpha_i = 0\}}, \tag{3.44}$$

where $Pr\{r_i | \alpha_i = 1\} = \sum_{\beta_i} Pr\{r_i | \alpha_i = 1, \beta_i\} P(\beta_i)$ and $Pr\{r_i | \alpha_i = 0\} = \sum_{\beta_i} Pr\{r_i | \alpha_i = 0, \beta_i\} P(\beta_i)$, with $\beta_i \in \{0, 1\}$. In the ideal case when β_i is known for $i = 1, \cdots, N_c$, the ML detector above can be further simplified. If we assume that $n_1, \cdots, n_{N_c}$ are i.i.d. circularly symmetric Gaussian random variables of zero-mean and variance σ_n^2, and $J_1, \cdots, J_{N_c}$ are i.i.d. circularly symmetric Gaussian random variables of zero-mean and variance $\sigma_{J_i}^2$, then it follows from (3.40) and (3.44) that:

$$\hat{k} = \arg \max_{1 \le i \le N_c} \frac{\|r_i\|^2 - \|r_i - s\|^2}{\sigma_i^2}, \tag{3.45}$$

where $\sigma_i^2 = \beta_i \sigma_{J_i}^2 + \sigma_n^2$.

Note that σ_i^2 is generally unknown. If we replace the overall interference power σ_i^2 with the instantaneous power of the received signal $\|r_i\|^2$, then it follows from (3.45) that:

$$\hat{k} = \arg \min_{1 \le i \le N_c} \frac{\|r_i - s\|^2}{\|r_i\|^2}. \tag{3.46}$$

For more tractable theoretical analysis, we can replace $\|r_i\|^2$ with the average signal power observed in channel i, $P_i = E\{\|r_i\|^2\}$. Define $Z_i \triangleq \frac{\|r_i - s\|}{\sqrt{P_i}}$, and then we have:

$$\hat{k} = \arg \min_{1 \le i \le N_c} Z_i. \tag{3.47}$$

Discussions In the conventional FH, a frequency synthesizer is used at the receiver to capture the transmitted signal. The strict requirement on frequency synchronization turns out to be a significant challenge in FH system design, especially for fast hopping systems. In MDFH and AJ-MDFH, a bandpass filter bank is used to capture the hopping frequency. The complexity is similar to using multiple FSK receivers in parallel. It is observed that (1) comparing with conventional FH, MDFH-based systems relax the frequency synchronization problem. At the same time, we show that if a bandpass filter bank is used by an adversary, then the conventional FH signal as well as the PN sequence can be easily captured, making it fragile to follower jamming and resulting in total loss of the transmission. In AJ-MDFH, the encrypted information is transmitted through hopping frequency control; this is like using the one-time key pad, making it impossible for the adversary to launch follower jamming. (2) Comparing with FSK, which has zero capacity under disguised jamming, AJ-MDFH is much more efficient and robust under disguised jamming.

3.4.4 Extension to Multi-carrier AJ-MDFH

For more efficient spectrum usage and robust jamming resistance, in this section, we extend the concept of MDFH to multi-carrier AJ-MDFH (MC-AJ-MDFH). The idea is to split all the N_c channels into N_g non-overlapping groups and each subcarrier hops within the assigned group based on the AJ-MDFH scheme. To ensure hopping randomness of all the subcarriers, the groups need to be reorganized or regenerated securely after a pre-specified period, named group period. A secure subgroup assignment algorithm can be developed as what we did in [24], to ensure that (i) each subcarrier hops over a new group of channels during each group period, so that it eventually hops over all the available channels in a pseudo-random manner, and (ii) only the legitimate receiver can recover the transmitted information correctly.

3.4.4.1 Multi-carrier AJ-MDFH without Diversity

In this case, each subcarrier transmits an independent bit stream. The spectral efficiency of the AJ-MDFH system can be increased significantly. Let $B_c = \log_2 N_c$ and $B_g = \log_2 N_g$, and then the number of bits transmitted by the MC-AJ-MDFH within each hopping period is $B_{MC} = (B_c - B_g)N_g = (B_c - \log_2 N_g)N_g$. B_{MC} is maximized when $B_g = B_c - 1$ or $B_g = B_c - 2$, which results in $B_{MC} = 2^{B_c - 1}$. Note that the number of bits transmitted by the AJ-MDFH within each hopping period is B_c; it can be seen that $B_{MC} > B_c$ as long as $B_c > 2$. Take $N_c = 256$ for example, and then the transmission efficiency of AJ-MDFH can be increased by $\frac{B_{MC}}{B_c} = \frac{2^{B_c-1}}{B_c} = 16$ times.

3.4.4.2 Multi-carrier AJ-MDFH with Diversity

Under multi-band jamming, diversity needs to be introduced to the AJ-MDFH system for robust jamming resistance. A natural solution to achieve frequency diversity is to transmit the same or correlated information through multiple subcarriers. The number of subcarriers needed to convey the same information varies in different jamming scenarios. Generally, the number of correlated signal subcarriers should not be less than the number of jammed bands. At the receiver, the received signals from different diversity branches can be combined for joint signal detection [25–27].

As will be shown in Sect. 3.4.7, MC-AJ-MDFH can increase the system efficiency and jamming resistance significantly through jamming randomization and frequency diversity. Moreover, by assigning different carrier groups to different users, MC-AJ-MDFH can also be used as a collision-free multiple access system.

3.4.5 *ID Constellation Design and Its Impact on System Performance*

For AJ-MDFH, ID signals are introduced to distinguish the true information channel from disguised channels invoked by jamming interference. In this section, we investigate ID constellation design and its impact on the performance of AJ-MDFH under various jamming scenarios.

3.4.5.1 Design Criterion and Jamming Classification

The general design criterion of the ID constellation is *to minimize the probability of error under a given signal power*. Under this criterion, the following questions need to be answered: (1) How does the size of the constellation impact the system performance? (2) How does the type or shape of the constellation influence the detection error? Which type should we use for optimal performance? In this section, we will try to address these questions under different jamming scenarios.

In literature, jamming has generally been modeled as Gaussian noise [28–30], referred to as *noise jamming*. Recall that disguised jamming denotes the jamming interference which has similar power and spectral characteristics as that of the true signal. For AJ-MDFH, when the ID constellation is known to, or can be guessed by the jammer, the jammer can then disguise itself by sending symbols taken from the same constellation over a different or fake channel. In this case, it could be difficult for the receiver to distinguish the true channel from the disguised channel, leading to high detection error probability. We refer to this kind of jamming as *ID jamming* or ID attack. ID jamming is the worst-case disguised jamming for AJ-MDFH.

3.4.5.2 Constellation Design Under Noise Jamming

Without loss of generality, we consider the case where the ID symbol is transmitted through channel 1, i.e.:

$$\boldsymbol{\alpha} = (\alpha_1, \cdots, \alpha_{N_c}) = (1, 0, \cdots, 0). \tag{3.48}$$

Recall that for $i = 1, \cdots, N_c$, $r_i = \alpha_i s + \beta_i J_i + n_i$. Let $\tilde{n}_i = \beta_i J_i + n_i$, and denote its variance as $\sigma_i^2 = \beta_i \sigma_{J_i}^2 + \sigma_n^2$, which may vary from channel to channel. Following the definition in (3.47), we have $Z_1 = \frac{\|\tilde{n}_1\|}{\sqrt{\|s\|^2+\sigma_1^2}}$, and $Z_i = \frac{\|\tilde{n}_i - s\|}{\sigma_i}$ for $2 \le i \le N_c$. It can be seen that Z_1 is a Rayleigh random variable with probability density function (PDF):

$$p_{Z_1}(z_1) = \frac{z_1}{\sigma^2} e^{-\frac{z_1^2}{2\sigma^2}}, \quad z_1 > 0, \tag{3.49}$$

where $\sigma^2 = \frac{\sigma_1^2}{2(\|s\|^2+\sigma_1^2)}$. For $2 \le i \le N_c$, Z_i is a Rician random variable with PDF:

$$p_{Z_i}(z_i) = \frac{z_i}{\sigma^2} e^{-\frac{z_i^2+\nu^2}{2\sigma^2}} I_0\left(\frac{z_i \nu}{\sigma^2}\right), \quad z_i > 0, \tag{3.50}$$

where $\nu = \frac{\|s\|}{\sigma_i}$, $\sigma = \frac{1}{\sqrt{2}}$ and $I_0(x)$ is the modified Bessel function of the first kind with order zero.

According to (3.47), the carrier can be correctly detected if and only if $Z_1 < Z_i$ for all $2 \leq i \leq N_c$. Assuming that the symbols in constellation Ω are equally probable, then the carrier detection error probability is given by:

$$P_e = 1 - \sum_{s \in \Omega} Pr\{Z_1 < Z_2, \ldots, Z_1 < Z_{N_c} | s\} p_S(s)$$

$$= 1 - \frac{1}{|\Omega|} \sum_{s \in \Omega} \int_0^\infty \prod_{i=2}^{N_c} Pr\{Z_i > z_1 | s, Z_1 = z_1\} p_{Z_1}(z_1) dz_1. \quad (3.51)$$

Note that $Z_2, \cdots, Z_{N_c}$ are i.i.d. Rician random variables, and then it follows from (3.49) and (3.50) that:

$$P_e = 1 - \frac{1}{|\Omega|} \sum_{s \in \Omega} \int_0^\infty \prod_{i=2}^{N_c} Q_1\left(\frac{\sqrt{2}}{\sigma_i}\|s\|, \sqrt{2}z_1\right) \frac{2(\|s\|^2 + \sigma_1^2)}{\sigma_1^2} z_1 e^{-\frac{\|s\|^2+\sigma_1^2}{\sigma_1^2} z_1^2} dz_1, \quad (3.52)$$

where Q_1 is the Marcum Q-function [18]. We have the following result:

Proposition 3.3 *Assuming the true channel index is k. Under noise jamming, an upper bound of the carrier detection error probability P_e can be obtained as:*

$$P_e^U = \frac{1}{|\Omega|} \sum_{s \in \Omega} \left[1 - \left(1 - \frac{\sigma_k^2}{\|s\|^2 + 2\sigma_k^2} e^{-\frac{\|s\|^2(\|s\|^2+\sigma_k^2)}{\sigma_m^2(\|s\|^2+2\sigma_k^2)}}\right)^{N_c-1}\right], \quad (3.53)$$

where $m = \arg\max\{\sigma_l^2\}$ *for* $1 \leq l \leq N_c, l \neq k$.

Proof It follows from (3.52) that:

$$P_e = 1 - \frac{1}{|\Omega|} \sum_{s \in \Omega} \int_0^\infty \prod_{l=2}^{N_c} Q_1\left(\frac{\sqrt{2}}{\sigma_l}\|s\|, \sqrt{2}z_1\right) p_{Z_1}(z_1) dz_1$$

$$\leq 1 - \frac{1}{|\Omega|} \sum_{s \in \Omega} \int_0^\infty Q_1^{N_c-1}\left(\frac{\sqrt{2}}{\sigma_m}\|s\|, \sqrt{2}z_1\right) p_{Z_1}(z_1) dz_1 \quad (3.54)$$

where $m = \arg\max_{2 \leq l \leq N_c}\{\sigma_l^2\}$. The inequality follows from the fact that for fixed $\|s\|$ and z_1, $Q_1\left(\frac{\sqrt{2}}{\sigma_l}\|s\|, \sqrt{2}z_1\right)$ is a monotonically decreasing function with respect to σ_l. The equality can be achieved when $\sigma_2 = \cdots = \sigma_{N_c}$.

Assume $N_c \geq 2$. For $N_c = 2$, it is easy to show that $P_e = P_e^U$ with $\sigma_m = \sigma_2$. Note that for fixed s and σ_m, $Q_1\left(\frac{\sqrt{2}}{\sigma_m}\|s\|, \sqrt{2}z_1\right) = Pr\{Z_m > z_1 | s, z_1 = Z_1\}$ is a function of z_1. And for $N_c > 2$, $f(x) = x^{N_c-1}$ is convex when $x > 0$. By Jensen's inequality, we obtain:

$$\int_0^\infty Q_1^{N_c-1}\left(\frac{\sqrt{2}}{\sigma_m}\|s\|, \sqrt{2}z_1\right) p_{Z_1}(z_1)dz_1$$
$$\geq \left[\int_0^\infty Pr\{Z_m > z_1|s, Z_1 = z_1\}p_{Z_1}(z_1)dz_1\right]^{N_c-1}$$
$$= [Pr\{Z_1 < Z_m|s\}]^{N_c-1}. \quad (3.55)$$

According to [31], $Pr\{Z_1 < Z_m|s\}$ can be calculated as:

$$Pr\{Z_1 < Z_m|s\} = 1 - \frac{\sigma_1^2}{\|s\|^2 + 2\sigma_1^2} e^{-\frac{\|s\|^2(\|s\|^2+\sigma_1^2)}{\sigma_m^2(\|s\|^2+2\sigma_1^2)}}. \quad (3.56)$$

Following (3.54), (3.55), and (3.56):

$$P_e \leq \frac{1}{|\Omega|}\sum_{s\in\Omega}\left[1 - \left(1 - \frac{\sigma_1^2}{\|s\|^2 + 2\sigma_1^2} e^{-\frac{\|s\|^2(\|s\|^2+\sigma_1^2)}{\sigma_m^2(\|s\|^2+2\sigma_1^2)}}\right)^{N_c-1}\right] = P_e^U \text{ for } N_c > 2. \quad (3.57)$$

Overall, $P_e \leq P_e^U$ for $N_c \geq 2$. ■

Assuming the true channel index is k, let $x = \frac{\|s\|^2}{\sigma_k^2}$ and $a(x) = 1 - (1 - \frac{1}{x+2}e^{-\zeta\frac{x^2+x}{x+2}})^{N_c-1}$. Then P_e^U can be written as $P_e^U = a(x)$ with $\zeta = \sigma_k^2/\sigma_m^2$. Note that when $x \gg 1$, $a(x) \approx \frac{(N_c-1)}{x+2}e^{-\zeta\frac{x^2+x}{x+2}} \triangleq \tilde{a}(x)$. It can be shown that when $x \gg 1$, $\tilde{a}(x)$ is a convex function. By Jensen's inequality [32], we have:

$$P_e^U \approx \frac{1}{|\Omega|}\sum_{s\in\Omega}\tilde{a}\left(\frac{\|s\|^2}{\sigma_k^2}\right) \geq \tilde{a}\left(\frac{1}{|\Omega|\sigma_k^2}\sum_{s\in\Omega}\|s\|^2\right) = \tilde{a}\left(\frac{P_s}{\sigma_k^2}\right). \quad (3.58)$$

The equality is achieved if and only if $\|s\|^2 = P_s$ for all $s \in \Omega$. This implies that *under the condition that the signal-to-jamming and noise ratio over channel k satisfies $\frac{\|s\|^2}{\sigma_k^2} \gg 1$, P_e^U is approximately minimized when the constellation is constant modulus, that is, $\|s\|^2 = P_s$ for all $s \in \Omega$.*

An intuitive explanation for this result is that the signal power in constant modulus constellations always equals to the maximal signal power available. Moreover, it can be seen that P_e^U is independent of the constellation size $|\Omega|$ but is only a function of P_s/σ_k^2. Next, we will investigate how the constellation size affects the system performance under ID attacks.

3.4.5.3 Constellation Design Under ID Jamming

Clearly, under ID attacks, the uncertainty of the ID symbol needs to be maximized. Under the assumption that all the symbols in a constellation Ω of size M are all equally probable, then the average symbol entropy:

$$H(s) = -\log_2 \frac{1}{|\Omega|} = \log_2 M \quad \text{bits}. \tag{3.59}$$

In the ideal case when the channel is noise-free, the optimal constellation size would be $M = \infty$. However, when noise is present, a larger M also implies there is a larger probability for an ID symbol to be mistaken for its neighboring symbols. More specifically, we have the following result:

Theorem 3.1 *For a given SNR and assuming PSK constellation is utilized, under ID jamming, the carrier detection error probability P_e is a function of constellation size M and*

$$\lim_{M\to\infty} P_e(M) = \bar{P}_e. \tag{3.60}$$

In other words, for any given $\epsilon > 0$, there always exists an M_t such that for all $M > M_t$, $|P_e(M) - \bar{P}_e| < \epsilon$.

Proof Note that the system is under ID attack. If the power of the M-ary PSK constellation is P_s, then the signal and jamming symbol can be written as $s = \sqrt{P_s}e^{j\frac{2\pi m_s}{M}}$, $J_j = \sqrt{P_s}e^{j\frac{2\pi m_J}{M}}$, respectively, where $M = |\Omega|$ and $0 \le m_s, m_J \le M-1$. Without loss of generality, we assume that: (i) Both the ID and jamming take the symbols in Ω with equal probability $1/M$; (ii) The signal is transmitted in channel 1 ($\alpha_1 = 1$) and channel j is jammed ($\beta_j = 1$).

- When $j = 1$, jamming collides with the ID signal. In this case, $r_1 = s + J_1 + n_1$ and $r_l = n_l$ for $l = 2, \ldots, N_c$. We have $Z_1 = \frac{\|J_1+n_1\|}{\sqrt{\|s+J_1\|^2+\sigma_n^2}}$ and $Z_l = \frac{\|n_l-s\|}{\sigma_n}$, where σ_n^2 is the noise variance. The detection error probability in this case can be calculated as:

$$P_{e1} = 1 - \frac{1}{M^2}\sum_{s\in\Omega}\sum_{J_1\in\Omega}\int_0^\infty [Pr\{z_1 < Z_2|s, J_1, z_1\}]^{N_c-1} p_{Z_1}(z_1)dz_1. \tag{3.61}$$

Note that Z_1 is a Rician random variable with PDF $p_{Z_1}(z_1) = \frac{z_1}{\sigma^2}e^{-\frac{z_1^2+\nu^2}{2\sigma^2}} I_0\left(\frac{z_1\nu}{\sigma^2}\right)$, where $\nu = \frac{\sqrt{P_s}}{\sqrt{\|s+J_1\|^2+\sigma_n^2}}$, $\sigma^2 = \frac{\sigma_n^2}{2(\|s+J_1\|^2+\sigma_n^2)}$ and Z_l's are i.i.d. Rician random variables with $\nu = \frac{\sqrt{P_s}}{\sigma_n}$, $\sigma^2 = \frac{1}{2}$. Then (3.61) can be written as:

$$
\begin{aligned}
P_{e1} =& 1 - \frac{1}{|\Omega|^2} \sum_{s\in\Omega} \sum_{J_1\in\Omega} \int_0^{\infty} Q_1^{N_c-1}\left(\frac{\sqrt{2P_s}}{\sigma_n}, \sqrt{2}z_1\right) 2z_1 \frac{\|s+J_1\|^2+\sigma_n^2}{\sigma_n^2} \\
&\cdot e^{-\frac{(\|s+J_1\|^2+\sigma_n^2)z_1^2+P_s}{\sigma_n^2}} I_0\left(\frac{2z_1}{\sigma_n^2}\sqrt{P_s(\|s+J_1\|^2+\sigma_n^2)}\right) dz_1 \\
=& 1 - \frac{1}{M}\sum_{\kappa=0}^{M-1} \int_0^{\infty} Q_1^{N_c-1}\left(\frac{\sqrt{2P_s}}{\sigma_n}, \sqrt{2}z_1\right) 2z_1 \left[\frac{2P_s}{\sigma_n^2}\left(1+\cos\frac{2\pi\kappa}{M}\right)+1\right] \\
&\cdot e^{-\left[\frac{2P_s}{\sigma_n^2}\left(1+\cos\frac{2\pi\kappa}{M}\right)+1\right]z_1^2-\frac{P_s}{\sigma_n^2}} I_0\left(\frac{2z_1}{\sigma_n^2}\sqrt{2P_s^2\left(1+\cos\frac{2\pi\kappa}{M}\right)+P_s\sigma_n^2}\right) dz_1,
\end{aligned}
\tag{3.62}
$$

where $\kappa \triangleq (m_s - m_J) \mod M$ is uniformly distributed over $[0, M-1]$.

- When $j = 2, \ldots, N_c$, jamming does not collide with ID signal. In this case, $r_1 = s + n_1$, $r_j = J_j + n_j$ and $r_l = n_l$ for $l = 2, \ldots, N_c, l \neq j$. We have $Z_1 = \frac{\|n_1\|}{\sqrt{P_s+\sigma_n^2}}$, $Z_j = \frac{\|J_j-s+n_j\|}{\sqrt{P_s+\sigma_n^2}}$ and $Z_l = \frac{\|n_l-s\|}{\sigma_n}$. The detection error probability in this case can be calculated as:

$$
\begin{aligned}
P_{e2} = 1 - \tfrac{1}{|\Omega|^2} \textstyle\sum_{s\in\Omega}\sum_{J_j\in\Omega}\int_0^{\infty} & Pr\{Z_j > z_1|s, J_j, z_1\} \\
& [Pr\{Z_l > z_1|s, z_1\}]^{N_c-2} p_{Z_1}(z_1)dz_1.
\end{aligned}
\tag{3.63}
$$

Note that Z_1 is a Rayleigh random variable with PDF $p_{Z_1}(z_1) = \frac{z_1}{\sigma^2}e^{-\frac{z_1^2}{2\sigma^2}}$, where $\sigma^2 = \frac{\sigma_n^2}{2(P_s+\sigma_n^2)}$, Z_j is a Rician random variable with $\nu = \frac{\|J_j-s\|}{\sqrt{P_s+\sigma_n^2}}$, $\sigma^2 = \frac{\sigma_n^2}{2(P_s+\sigma_n^2)}$ and Z_l's are i.i.d. Rician random variables with $\nu = \frac{\sqrt{P_s}}{\sigma_n}$, $\sigma^2 = \frac{1}{2}$. Then (3.63) can be written as:

$$
\begin{aligned}
P_{e2} =& 1 - \frac{1}{|\Omega|^2} \sum_{s\in\Omega}\sum_{J_1\in\Omega}\int_0^{\infty} Q_1\left(\frac{\sqrt{2}}{\sigma_n}\|J_j - s\|, \frac{\sqrt{2(P_s+\sigma_n^2)}}{\sigma_n}z_1\right) \\
&\cdot Q_1^{N_c-2}\left(\frac{\sqrt{2P_s}}{\sigma_n}, \sqrt{2}z_1\right) \frac{2(P_s+\sigma_n^2)}{\sigma_n^2} z_1 e^{-\frac{P_s+\sigma_n^2}{\sigma_n^2}z_1^2} dz_1 \\
=& 1 - \frac{1}{M}\sum_{\kappa=0}^{M-1}\int_0^{\infty} Q_1\left(\frac{2}{\sigma_n}\sqrt{P_s\left(1-\cos\frac{2\pi\kappa}{M}\right)}, \frac{1}{\sigma_n}\sqrt{2(P_s+\sigma_n^2)}z_1\right) \\
&\cdot Q_1^{N_c-2}\left(\frac{\sqrt{2P_s}}{\sigma_n}, \sqrt{2}z_1\right) \frac{2(P_s+\sigma_n^2)}{\sigma_n^2} z_1 e^{-\frac{P_s+\sigma_n^2}{\sigma_n^2}z_1^2} dz_1.
\end{aligned}
\tag{3.64}
$$

The overall detection error probability in noisy environment is given as:

$$P_e = Pr\{j = 1\}P_{e1} + Pr\{2 \leq j \leq N_c\}P_{e2} = \frac{1}{N_c}P_{e1} + \frac{N_c - 1}{N_c}P_{e2}. \tag{3.65}$$

When $\frac{P_s}{\sigma_n^2}$ is fixed, it follows from (3.62) and (3.64) that P_e is a function of M given as:

$$P_e = \frac{1}{M}\sum_{\kappa=0}^{M-1} b\left(\frac{2\pi\kappa}{M}\right), \tag{3.66}$$

where:

$$\begin{aligned} b(x) = 1 &- \frac{1}{N_c}\int_0^{\infty} Q_1^{N_c-1}\left(\frac{\sqrt{2P_s}}{\sigma_n}, \sqrt{2}z_1\right) 2z_1 \left[\frac{2P_s}{\sigma_n^2}(1+\cos x)+1\right] \\ &\cdot e^{-\left[\frac{2P_s}{\sigma_n^2}(1+\cos x)+1\right]z_1^2 - \frac{P_s}{\sigma_n^2}} I_0\left(\frac{2z_1}{\sigma_n^2}\sqrt{2P_s^2(1+\cos x)+P_s\sigma_n^2}\right) dz_1 \\ &- \frac{N_c-1}{N_c}\int_0^{\infty} Q_1\left(\frac{2}{\sigma_n}\sqrt{P_s(1-\cos x)}, \frac{1}{\sigma_n}\sqrt{2(P_s+\sigma_n^2)}z_1\right) \\ &\cdot Q_1^{N_c-2}\left(\frac{\sqrt{2P_s}}{\sigma_n}, \sqrt{2}z_1\right) \frac{2(P_s+\sigma_n^2)}{\sigma_n^2} z_1 e^{-\frac{P_s+\sigma_n^2}{\sigma_n^2}z_1^2} dz_1. \end{aligned} \tag{3.67}$$

As M approaches infinity, P_e converges. In fact, we have:

$$\begin{aligned} \bar{P_e} = \lim_{M\to\infty} P_e &= \lim_{M\to\infty} \sum_{\kappa=0}^{M-1} \frac{b\left(\frac{2\pi\kappa}{M}\right)\cdot\frac{2\pi}{M}}{M\cdot\frac{2\pi}{M}} \\ &= \frac{1}{2\pi}\lim_{M\to\infty}\sum_{\kappa=0}^{M-1} b\left(\frac{2\pi\kappa}{M}\right)\cdot\frac{2\pi}{M} \\ &= \frac{1}{2\pi}\int_0^{2\pi} b(x)dx. \end{aligned} \tag{3.68}$$

Note that $b(x)$ is the detection error probability when the angle between the signal symbol and the jamming symbol is x; hence $0 \leq b(x) \leq 1$ and we have:

$$0 \leq \bar{P}_e = \frac{1}{2\pi}\int_0^{2\pi} b(x)dx \leq 1. \tag{3.69}$$

That is: $\forall\epsilon > 0$, there always exists an integer M_t such that $\forall M > M_t$, $|P_e - \bar{P}_e| < \epsilon$.

■

Remark 3.2 This theorem essentially says that for a given SNR, due to the noise effect, increasing the constellation size over a threshold M_t will result in little improvement in detection error probability. *This result justifies the use of finite constellation in AJ-MDFH.*

3.4.6 Spectral Efficiency Analysis of AJ-MDFH

The spectral efficiency ν is defined as the ratio of the information bit rate R_b to the transmission bandwidth W_t, i.e., $\nu = \frac{R_b}{W_t}$. In this section, we will analyze and compare the spectral efficiency of the conventional and the new frequency hopping schemes, including conventional FH, MDFH, and (MC-)AJ-MDFH.

We start with the *single-user case*. Recall that T_s and T_h denote the symbol period and the hopping duration, respectively; $N_h = T_s/T_h$ is the number of hops per symbol period. For fair comparison, we assume that all systems have (i) the same number of available channels N_c, (ii) the same hopping period T_h to ensure the hopping channels have the same bandwidth $W_c = 2/T_h$, and (iii) the same frequency spacing Δf between two adjacent subcarriers, where $\Delta f \geq 1/T_h$ is chosen to avoid inter-carrier interference. Note that under these assumptions, all systems have the same total bandwidth $W_t = (N_c - 1)\Delta f + W_c$. For conventional FH using MFSK modulation, $\log_2 M$ bits are transmitted during each symbol period. The bit rate of conventional FH can be calculated as $R_b = \frac{\log_2 M}{T_s} = \frac{\log_2 M}{T_h N_h}$, and the corresponding spectral efficiency can be obtained as $\nu = \frac{R_b}{W_t} = \frac{\log_2 M}{T_h N_h W_t}$. The bit rate and spectral efficiency of other frequency hopping schemes can be obtained similarly. The results are listed in Table 3.2.

Next, we consider the more general *multiuser case*. The multiple access scheme for conventional FH, namely, FHMA, was proposed in [2]. The multiple access extension of MDFH, denoted as E-MDFH, has been analyzed in [15]. Due to the variability in multiple access system design, a closed-form expression of the spectral efficiency is hard to obtain. Here we compare the total information bits allowed to be transmitted by each system under the same BER and bandwidth requirements and illustrate the spectral efficiency comparison through the following example.

Let N_u denote the number of users and we choose the number of channels be $N_c = 64$ (i.e., $B_c = 6$). For MC-AJ-MDFH, we choose $N_g = N_u = 4$. For E-MDFH, we choose 8-PSK to modulate $B_s = 3$ ordinary bits and $N_g = N_u = 4$, $N_h = 3$. For FHMA, we choose 64-FSK modulation, $N_h = 3$, and consider $N_u = 2, 3, 4$, respectively. The required BER is 10^{-4}. Figure 3.8 depicts the performance

Table 3.2 The bit rate (R_b) and spectral efficiency (ν) comparison in the single-user case. (© [2013] IEEE. Reprinted, with permission, from Ref. [21])

	Conventional FH	MDFH	AJ-MDFH	MC-AJ-MDFH
R_b (b/s)	$\frac{\log_2 M}{T_h N_h}$	$\frac{N_h B_c + B_s}{N_h T_h}$	$\frac{B_c}{T_h}$	$\frac{(B_c - \log_2 N_g) N_g}{T_h}$
ν $(b/s/Hz)$	$\frac{\log_2 M}{T_h N_h W_t}$	$\frac{N_h B_c + B_s}{T_h N_h W_t}$	$\frac{B_c}{T_h W_t}$	$\frac{(B_c - \log_2 N_g) N_g}{T_h W_t}$

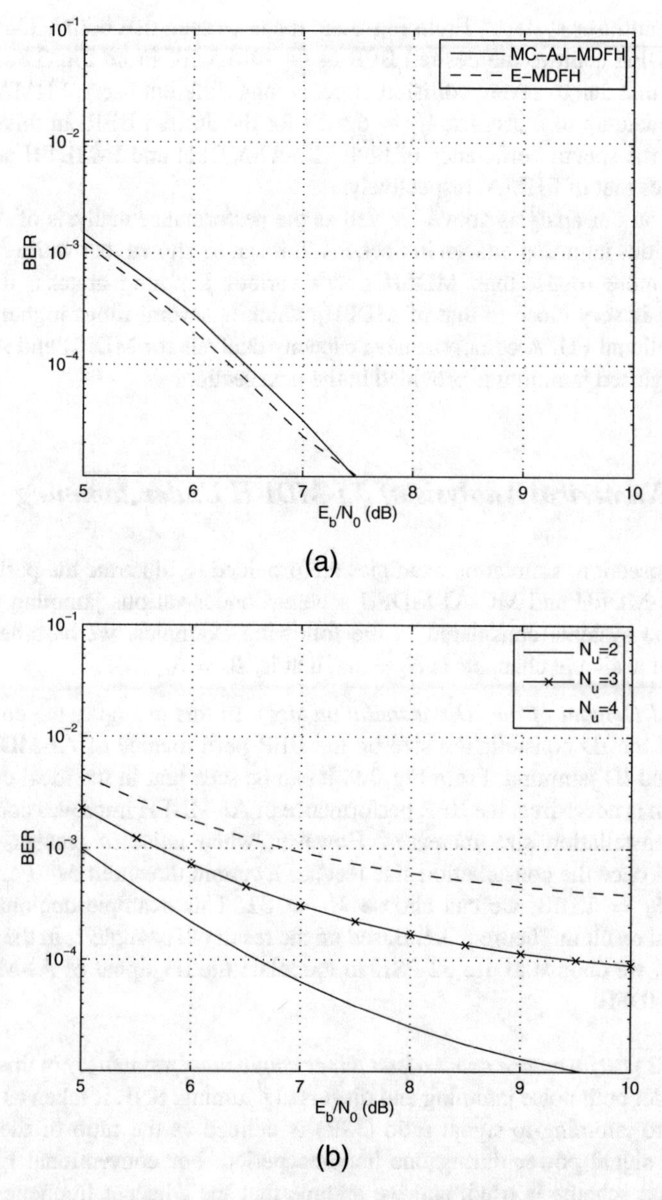

Fig. 3.8 Performance comparison of MC-AJ-MDFH, E-MDFH, and FHMA in the multiuser case. Here N_u denotes the number of users in the system. (**a**) E-MDFH and MC-AJ-MDFH, $N_u = 4$. (**b**) FHMA. (© [2013] IEEE. Reprinted, with permission, from Ref. [21])

of these multiuser systems. From Fig. 3.8a, it can be seen that both MC-AJ-MDFH and E-MDFH achieve the desired BER at $\frac{E_b}{N_0} \approx 6.5$ dB. From Fig. 3.8b, it can be observed that due to severe collision effect among different users, FHMA can only accommodate up to 2 users at $\frac{E_b}{N_0} \approx 6.5$ dB for the desired BER. In this particular example, the spectral efficiency of both MC-AJ-MDFH and E-MDFH are 4 times and 5 times that of FHMA, respectively.

Based on our analysis above, as well as the performance analysis of AJ-MDFH under various jamming attacks in Sect. 3.4.7, it can be shown that while AJ-MDFH is much more robust than MDFH under various jamming attacks, its spectral efficiency is very close to that of MDFH, which is several times higher than that of conventional FH. A comprehensive capacity analysis for MDFH and AJ-MDFH under disguised jamming is provided in the next section.

3.4.7 Numerical Analysis of AJ-MDFH Under Jamming

In this subsection, simulation examples are provided to illustrate the performances of the AJ-MDFH and MC-AJ-MDFH schemes under various jamming scenarios. For all the systems considered in the following examples, we assume the total number of available channels is $N_c = 64$, that is, $B_c = 6$.

Example 1 (Impact of the ID constellation size) In this example, we consider the impact of the ID constellation size on the BER performance of AJ-MDFH under single band ID jamming. From Fig. 3.9, it can be seen that in the ideal case where the system is noise-free, the BER performance of AJ-MDFH improves continuously as the constellation size increases. However, when noise is present, the BER converges once the constellation size reaches a certain threshold M_t. For example, for $E_b/N_0 = 15$ dB, we can choose $M_t = 32$. This example demonstrates the theoretical result in Theorem 3.1. Based on the result of Example 1, in the following examples, we choose to use 32-PSK to modulate the ID signal of AJ-MDFH and MC-AJ-MDFH.

Example 2 (Performance comparison under single band jamming) In this example, we consider both noise jamming and disguised jamming. SNR is taken as $E_b/N_0 = 10$ dB, and jamming-to-signal ratio (JSR) is defined as the ratio of the jamming power to signal power during one hopping period. For conventional FH, 4-FSK modulation scheme is used, and we assume that the adjacent frequency tones in 4-FSK correspond to the center frequencies of the adjacent MDFH channels. For MDFH, QPSK is used to modulate ordinary bits, and the number of hops per symbol period is $N_h = 3$. From Fig. 3.10, it can be observed that AJ-MDFH can effectively reduce the performance degradation caused by disguised jamming, while remaining robust under noise jamming. Note that JSR = 0 dB under disguised jamming corresponds to the ID jamming for AJ-MDFH. It can also be seen that the performance of AJ-MDFH improves significantly when the jamming power differs

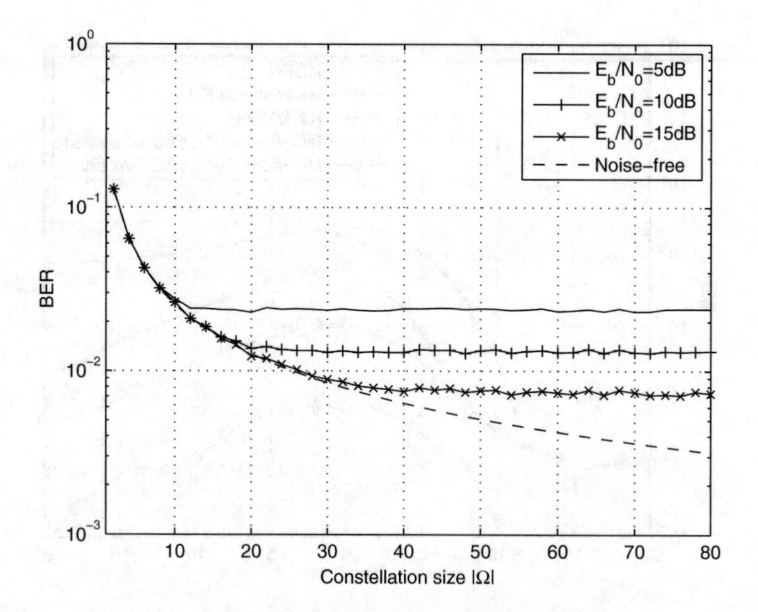

Fig. 3.9 Example 1: The performance of AJ-MDFH with different constellation size, under single band ID jamming. (© [2013] IEEE. Reprinted, with permission, from Ref. [21])

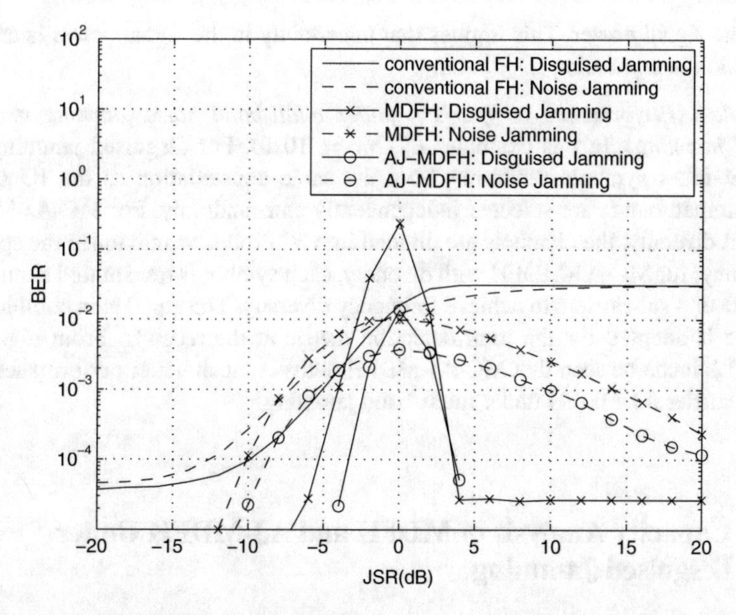

Fig. 3.10 Example 2: Performance comparison under single band jamming. (© [2013] IEEE. Reprinted, with permission, from Ref. [21])

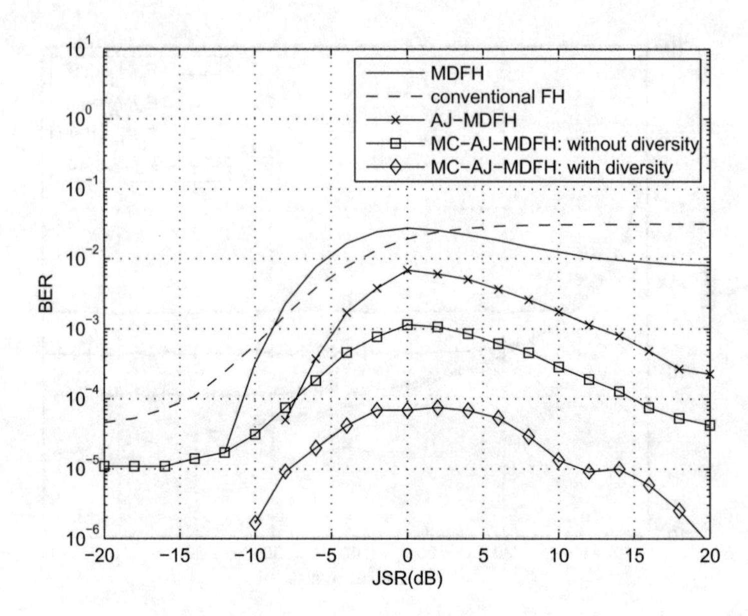

Fig. 3.11 Example 3: Performance comparison under 2-band noise jamming. (© [2013] IEEE. Reprinted, with permission, from Ref. [21])

from the signal power. This implies that uncertainty in the signal power is another dimension in combating ID jamming.

Example 3 (Performance comparison under multi-band noise jamming and disguised jamming) In this example, $E_b/N_0 = 10\,\text{dB}$. For disguised jamming, the jammer takes symbols randomly from the same constellation as the ID signal. The jammed bands are selected independently and randomly. For MC-AJ-MDFH without diversity, the channels are divided into 32 groups to maximize the spectral efficiency; for MC-AJ-MDFH with diversity, each symbol is transmitted simultaneously over 4 subcarriers to achieve frequency diversity. The equal gain combination scheme is adopted for the joint detection metric at the receiver. From Figs. 3.11 and 3.12, it can be seen that MC-AJ-MDFH delivers much better performance than single carrier AJ-MDFH under multi-band jamming.

3.5 Capacity Analysis of MDFH and AJ-MDFH Under Disguised Jamming

In this section, we analyze the capacity of MDFH and AJ-MDFH under disguised jamming. Both MDFH and AJ-MDFH are modeled as arbitrarily varying channels (AVCs) [33–38], which is characterized as $W : \mathcal{X} \times \mathcal{J} \rightarrow \mathcal{S}$, where $\mathcal{X}$ is the

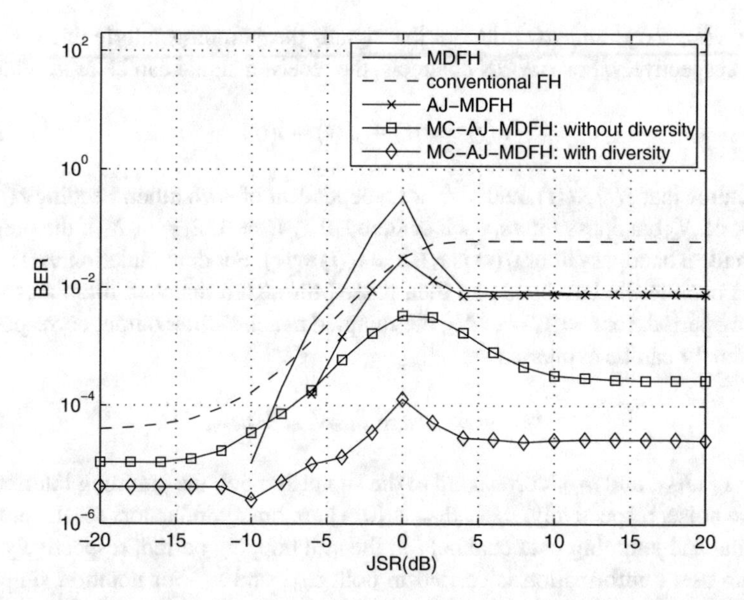

Fig. 3.12 Example 3: Performance comparison under 2-band disguised jamming. (© [2013] IEEE. Reprinted, with permission, from Ref. [21])

transmitted signal space, $\mathcal{J}$ is the jamming space, and $\mathcal{S}$ is the estimated information space. For any $\mathbf{x} \in \mathcal{X}$, $\mathbf{J} \in \mathcal{J}$, and $\mathbf{s} \in \mathcal{S}$, $W(\mathbf{s}|\mathbf{x}, \mathbf{J})$ denotes the conditional probability that $\mathbf{s}$ is detected at the receiver, given that $\mathbf{x}$ is the transmitted signal and $\mathbf{J}$ is the jamming. If $\mathcal{J} = \mathcal{X}$ and $W(\mathbf{s}|\mathbf{x}, \mathbf{J}) = W(\mathbf{s}|\mathbf{J}, \mathbf{x})$ for any $\mathbf{x}, \mathbf{J} \in \mathcal{X}, \mathbf{s} \in \mathcal{S}$, the AVC is said to have a *symmetric kernel* [39]. Define $\hat{W} : \mathcal{X} \times \mathcal{X} \rightarrow \mathcal{S}$ by $\hat{W}(\mathbf{s}|\mathbf{x}, \mathbf{J}) \triangleq \sum_{\mathbf{y} \in \mathcal{Y}} \pi(\mathbf{y}|\mathbf{J}) W(\mathbf{s}|\mathbf{x}, \mathbf{y})$, where $\pi : \mathcal{X} \rightarrow \mathcal{Y}$ is a probability matrix and $\mathcal{Y} \subseteq \mathcal{J}$. If there exists a π such that $\hat{W}(\mathbf{s}|\mathbf{x}, \mathbf{J}) = \hat{W}(\mathbf{s}|\mathbf{J}, \mathbf{x}), \;\; \forall \mathbf{x}, \mathbf{J} \in \mathcal{X}, \;\; \forall \mathbf{s} \in \mathcal{S}$, then W is said to be *symmetrizable*. The deterministic code[2] capacity of an AVC for the average probability of error is positive iff the AVC is nonsymmetrizable [36, 37, 39, 40].

3.5.1 Capacity of MDFH Under Disguised Jamming

In this section, we will show that for MDFH, due to the fact that there is no shared secret between the transmitter and the receiver, when the constellation Ω and the pulse shaping filter $g(t)$ are known to the jammer, the capacity of MDFH under the worst-case disguised jamming is actually zero.

[2] A deterministic (n, k) code means that each k-bit data word is mapped to a unique n-bit codeword.

Let $s(t)$, $J(t)$, and $n(t)$ denote the signal, the jamming interference, and the noise, respectively. For AWGN channels, the received signal can be represented as:

$$r(t) = s(t) + J(t) + n(t). \tag{3.70}$$

We assume that $s(t)$, $J(t)$, and $n(t)$ are independent of each other. Feeding $r(t)$ into a bank of N_c bandpass filters, each centered at f_i $(i = 1, 2, \cdots, N_c)$, the output of the ith ideal bandpass filter $\mathrm{f}_i(t)$ is $r_i(t) = \mathrm{f}_i(t) * r(t)$. For demodulation, $r_i(t)$ is first shifted back to the baseband and then passed through a matched filter. At the mth hopping period, for $i = 1, \cdots, N_c$, the sampled matched filter output corresponding to channel i can be expressed as:

$$r_{i,m} = \alpha_{i,m} s_n + \beta_{i,m} J_{i,m} + n_{i,m}, \tag{3.71}$$

where s_n, $J_{i,m}$, and $n_{i,m}$ correspond to the signal symbol, the jamming interference, and the noise, respectively; $\alpha_{i,m}, \beta_{i,m} \in \{0, 1\}$ are binary indicators for the presence of signal and jamming over channel i at the mth hopping period, respectively. Note that the user's information is carried in both $\alpha_{i,m}$ and s_n. For notation simplicity, without loss of generality, we omit the subscript m and n in (3.71). That is, for a particular hopping period, (3.71) is reduced to:

$$r_i = \alpha_i s + \beta_i J_i + n_i, \quad i = 1, \cdots, N_c. \tag{3.72}$$

The carrier bits and the ordinary bits can then be estimated from r_i [21].

Following (3.72), let $\mathbf{r} = (r_1, \ldots, r_{N_c})$, $\boldsymbol{\alpha} = (\alpha_1, \ldots, \alpha_{N_c})$, $\boldsymbol{\beta} = (\beta_1, \ldots, \beta_{N_c})$, $\mathbf{J}' = (J_1, \cdots, J_{N_c})$ and $\mathbf{n} = (n_1, \ldots, n_{N_c})$, the MDFH system under hostile jamming can be modeled as:

$$\mathbf{r} = \boldsymbol{\alpha} s + \boldsymbol{\beta} \cdot \mathbf{J}' + \mathbf{n}. \tag{3.73}$$

Note that in MDFH, the information is contained in both $\boldsymbol{\alpha}$ and s. Define $\mathbf{x} = \boldsymbol{\alpha} s$, $\mathbf{J} = \boldsymbol{\beta} \cdot \mathbf{J}'$, and then we have:

$$\mathbf{r} = \mathbf{x} + \mathbf{J} + \mathbf{n}. \tag{3.74}$$

Let $\mathcal{A} = \{\boldsymbol{\alpha} = (\alpha_1, \ldots, \alpha_{N_c}) | \alpha_i \text{ is 0 or 1, and } \sum_{i=1}^{N_c} \alpha_i = 1\}$. Define $\mathcal{X} = \{\boldsymbol{\alpha} s | \boldsymbol{\alpha} \in \mathcal{A}, s \in \Omega\}$ be the set of all possible information signal $\mathbf{x}$, and $\mathcal{J} = \{\mathbf{J} = (\beta_1 J_1, \cdots, \beta_{N_c} J_{N_c}) | J_i \in \Omega_J, \beta_i = 0 \text{ or } 1, i = 1, \cdots, N_c\}$, where Ω_J is the set of all possible jamming symbols. Let $\hat{\mathbf{x}}$ be the estimated version of $\mathbf{x}$ at the receiver and $W_0(\hat{\mathbf{x}}|\mathbf{x}, \mathbf{J})$ the conditional probability that $\hat{\mathbf{x}}$ is estimated at the receiver given that the signal is $\mathbf{x}$ and the jamming is $\mathbf{J}$. The jammed MDFH system can be modeled as an arbitrarily varying channel (AVC) characterized by the probability matrix:

$$W_0 : \mathcal{X} \times \mathcal{J} \rightarrow \mathcal{X}, \tag{3.75}$$

with:

$$W_0(\hat{\mathbf{x}}|\mathbf{x}, \mathbf{J}) \geq 0, \quad \hat{\mathbf{x}}, \mathbf{x} \in \mathcal{X}, \mathbf{J} \in \mathcal{J}, \tag{3.76}$$

$$\sum_{\hat{\mathbf{x}} \in \mathcal{X}} W_0(\hat{\mathbf{x}}|\mathbf{x}, \mathbf{J}) = 1, \; \mathbf{x} \in \mathcal{X}, \mathbf{J} \in \mathcal{J}. \tag{3.77}$$

W_0 is called the kernel of the AVC.

Under the worst-case single band disguised jamming:

$$\mathcal{J} = \{\boldsymbol{\beta} b | \boldsymbol{\beta} \in \mathcal{A}, b \in \Omega\} = \mathcal{X}. \tag{3.78}$$

That is, the jamming and the information signal are fully symmetric. Note that in MDFH, no shared randomness is exploited for signal detection at the receiver, the recovery of $\mathbf{x}$ is fully based on $\mathbf{r}$, and we further have:

$$W_0(\hat{\mathbf{x}}|\mathbf{x}, \mathbf{J}) = W_0(\hat{\mathbf{x}}|\mathbf{J}, \mathbf{x}). \tag{3.79}$$

This implies that the kernel of the AVC corresponding to MDFH, W_0, is symmetric.

In [39], it has been proved that the deterministic capacity (i.e., the largest rate achieved with deterministic codes) of an AVC with symmetric kernel is zero. Therefore, we have the result below.

Proposition 3.4 *The deterministic capacity of MDFH under the worst-case single band disguised jamming is zero.*

3.5.2 Capacity of AJ-MDFH Under Disguised Jamming

In this section, first, we will show that due to the shared randomness introduced by the secure ID sequence, the AVC kernel corresponding to AJ-MDFH is nonsymmetrizable even under the worst-case disguised jamming—ID jamming. We will further derive the capacity of AJ-MDFH under ID jamming.

In AJ-MDFH, each user is assigned a secure ID sequence. For each hopping period, AJ-MDFH can also be characterized by (3.72), except that s is now the ID symbol instead of the signal symbol. *It should be noted that to prevent impersonate ID attack, the ID symbol is refreshed at each hopping period.* For AJ-MDFH, the user's information is only carried in α_i. Recall that for AJ-MDFH, the ML receiver reduces to a normalized minimum distance receiver [21]. Define $P_i = E\{\|r_i\|^2\}$ and:

$$Z_i = \frac{\|r_i - s\|}{\sqrt{P_i}}. \tag{3.80}$$

Let $\mathcal{I}_c = \{1, \cdots, N_c\}$. Assuming $\alpha_i = \delta(k-i)$ for some $k \in \mathcal{I}_c$, at the receiver, k is then estimated as $\hat{k} = \arg\min_{i \in \mathcal{I}_c} Z_i$.

3.5.2.1 AVC Symmetricity Analysis

Recall that for AJ-MDFH:

$$\mathbf{r} = \boldsymbol{\alpha} s + \mathbf{J} + \mathbf{n}. \tag{3.81}$$

Under the worst-case single band disguised jamming, $\mathbf{J} \in \mathcal{X}$, and can be represented as $\mathbf{J} = \boldsymbol{\beta} b$ for some $\boldsymbol{\beta} \in \mathcal{A}$ and $b \in \Omega$. Note that the information is only transmitted through $\boldsymbol{\alpha}$; the AVC corresponding to AJ-MDFH can be characterized by the probability matrix:

$$W : \mathcal{X} \times \mathcal{X} \to \mathcal{A}, \tag{3.82}$$

with:

$$\begin{aligned} & W(\hat{\boldsymbol{\alpha}}|\mathbf{x}, \mathbf{J}) \geq 0, \mathbf{x} = \boldsymbol{\alpha} s \in \mathcal{X}, \mathbf{J} = \boldsymbol{\beta} b \in \mathcal{X}, \hat{\boldsymbol{\alpha}}, \boldsymbol{\alpha}, \boldsymbol{\beta} \in \mathcal{A}, \\ & \sum_{\hat{\boldsymbol{\alpha}} \in \mathcal{A}} W(\hat{\boldsymbol{\alpha}}|\mathbf{x}, \mathbf{J}) = 1, \forall\ (\mathbf{x}, \mathbf{J}) \in \mathcal{X}^2. \end{aligned} \tag{3.83}$$

Here $\hat{\boldsymbol{\alpha}}$ is the estimated version of $\boldsymbol{\alpha}$. In this section, we will first prove that *under reasonable SNR levels, the kernel W defined in* (3.82) *and* (3.83) *is nonsymmetric*. Then prove a stronger result: *W is actually nonsymmetrizable*.

For AJ-MDFH, W is symmetric if and only if:

$$W(\hat{\boldsymbol{\alpha}}|\mathbf{x}, \mathbf{J}) = W(\hat{\boldsymbol{\alpha}}|\mathbf{J}, \mathbf{x}),\ \ \forall \mathbf{x}, \mathbf{J} \in \mathcal{X},\ \ \forall \hat{\boldsymbol{\alpha}} \in \mathcal{A}. \tag{3.84}$$

To prove that W is nonsymmetric, we need to show that there always exists some $\mathbf{x}$, $\mathbf{J}$ and $\hat{\boldsymbol{\alpha}}$, such that the equality above does not hold. Following the discussion on ID constellation design in Part I [21], we assume that Ω is a PSK constellation with power P_s and define a mapping $\upsilon : \mathcal{I}_c \to \mathcal{A}$ as:

$$\upsilon(k) = \boldsymbol{\alpha} \text{ if } \alpha_i = \delta(k-i), \forall i \in \mathcal{I}_c. \tag{3.85}$$

Lemma 3.1 *Suppose* X, Y *are independent continuous random variables. If* $Z_1, \cdots, Z_N$ *are i.i.d. continuous random variables, which are also independent of* X, Y*, then:*

$$Pr\{X < Y \text{ and } X < Z_i, \forall 1 \leq i \leq N\} \geq Pr\{X < Y\} - N\, Pr\{X \geq Z_{i_0}\}, \tag{3.86}$$

for any fixed $1 \leq i_0 \leq N$.

Proof Let $\mathcal{A} = \{X < Y\}$, $\mathcal{B} = \{X < Z_i, \forall 1 \leq i \leq N\}$, and $\bar{\mathcal{B}}$ be the complement of $\mathcal{B}$. Then the inequality (3.86) follows from the fact that $Pr\{\bar{\mathcal{B}}\} \geq \sum_{i=1}^{N} Pr\{X \geq Z_i\} = NPr\{X \geq Z_{i_0}\}$ for any fixed $1 \leq i_0 \leq N$, and $Pr\{\mathcal{A} \bigcap \mathcal{B}\} = Pr\{\mathcal{A}\} - Pr\{\mathcal{A} \bigcap \bar{\mathcal{B}}\} \geq Pr\{\mathcal{A}\} - Pr\{\bar{\mathcal{B}}\}$. ■

Proposition 3.5 *Assuming Ω is a PSK constellation with power P_s, let $\mathbf{x} = \boldsymbol{\alpha} s$, $\mathbf{J} = \boldsymbol{\beta} b$, where $\boldsymbol{\alpha}, \boldsymbol{\beta} \in \mathcal{A}$, $\boldsymbol{\alpha} \neq \boldsymbol{\beta}$, and $s, b \in \Omega$, then:*

$$W(\boldsymbol{\alpha}|\mathbf{x}, \mathbf{J}) \geq 1 - \frac{1}{2} e^{-\frac{\|b-s\|^2}{2\sigma_n^2}} - \epsilon, \tag{3.87}$$

where $\epsilon = \frac{N_c - 2}{\gamma + 2} \exp\{-\frac{\gamma(\gamma+1)}{\gamma+2}\}$ with $\gamma = \frac{P_s}{\sigma_n^2}$ denoting the SNR.

Proof Let $\boldsymbol{\alpha} = v(k)$ and $\boldsymbol{\beta} = v(j)$. When $\boldsymbol{\beta} \neq \boldsymbol{\alpha}$, we have $j \neq k$ and:

$$\begin{aligned} W(\boldsymbol{\alpha}|\mathbf{x}, \mathbf{J}) &= W(\boldsymbol{\alpha}|\boldsymbol{\alpha} s, \boldsymbol{\beta} b) \\ &= Pr\{Z_k < Z_j \text{ and } Z_k < Z_i, \forall i \in \mathcal{I}_c, i \neq j, k|\mathbf{x}, \mathbf{J}\}. \end{aligned} \tag{3.88}$$

From Lemma 3.1, we have for any fixed $i_0 \in \mathcal{I}_c, i_0 \neq k, j$:

$$W(\boldsymbol{\alpha}|\mathbf{x}, \mathbf{J}) \geq Pr\{Z_k < Z_j|\mathbf{x}, \mathbf{J}\} - (N_c - 2)Pr\{Z_k \geq Z_{i_0}|\mathbf{x}, \mathbf{J}\}. \tag{3.89}$$

For any $i \in \mathcal{I}_c$, it follows from (3.72) and (3.80) that the received signal r_i and the corresponding metric Z_i can be written as:

$$r_i = \begin{cases} s + n_k, & i = k, \\ b + n_j, & i = j, \\ n_i, & i \neq j, k, \end{cases} \quad Z_i = \begin{cases} \frac{\|n_k\|}{\sqrt{P_s + \sigma_n^2}}, & i = k, \\ \frac{\|b - s + n_j\|}{\sqrt{P_s + \sigma_n^2}}, & i = j, \\ \frac{\|n_i - s\|}{\sigma_n}, & i \neq j, k. \end{cases} \tag{3.90}$$

Then, $Pr\{Z_k < Z_j|\mathbf{x}, \mathbf{J}\} = Pr\{\|n_k\| < \|b - s + n_j\||\mathbf{x}, \mathbf{J}\}$. For any $s, b \in \Omega$, both n_k and $b - s + n_j$ are circularly symmetric complex Gaussian random variables with $n_k \sim \mathcal{CN}(0, \sigma_n^2)$ and $b - s + n_j \sim \mathcal{CN}(b - s, \sigma_n^2)$. Then $Pr\{Z_k < Z_j|\mathbf{x}, \mathbf{J}\}$ can be calculated as (see [31], page 49):

$$Pr\{Z_k < Z_j|\mathbf{x}, \mathbf{J}\} = 1 - \frac{1}{2} e^{-\frac{\|b-s\|^2}{2\sigma_n^2}}. \tag{3.91}$$

Similarly, for any fixed $i_0 \in \mathcal{I}_c, i_0 \neq k, j$, we have $\frac{n_k}{\sqrt{P_s + \sigma_n^2}} \sim \mathcal{CN}(0, \frac{\sigma_n^2}{P_s + \sigma_n^2})$, $\frac{n_{i_0} - s}{\sigma_n} \sim \mathcal{CN}(-\frac{s}{\sigma_n}, 1)$ and:

$$\begin{aligned} Pr\{Z_k \geq Z_{i_0}|\mathbf{x}, \mathbf{J}\} &= Pr\{\frac{\|n_k\|}{\sqrt{P_s + \sigma_n^2}} \geq \frac{\|n_{i_0} - s\|}{\sigma_n}\} \\ &= \frac{1}{\gamma + 2} e^{-\frac{\gamma(\gamma+1)}{\gamma+2}}, \end{aligned} \tag{3.92}$$

where $\gamma = \frac{P_s}{\sigma_n^2}$. It then follows from (3.89), (3.90), (3.91), and (3.92) that:

$$W(\boldsymbol{\alpha}|\mathbf{x}, \mathbf{J}) \geq 1 - \frac{1}{2}e^{-\frac{\|b-s\|^2}{2\sigma_n^2}} - \epsilon. \tag{3.93}$$

■

Note that ϵ is determined by the SNR γ as well as the number of channels N_c. When $SNR \geq 10\,\text{dB}$ and $N_c = 512$, for example, $\epsilon \leq 0.004$.

Theorem 3.2 *Assuming Ω is a PSK constellation with power P_s, let $\mathbf{x} = \boldsymbol{\alpha} s$, $\mathbf{J} = \boldsymbol{\beta} b$, where $\boldsymbol{\alpha}, \boldsymbol{\beta} \in \mathcal{A}$, $\boldsymbol{\alpha} \neq \boldsymbol{\beta}$, and $s, b \in \Omega$, $s \neq b$. Let $\gamma = \frac{P_s}{\sigma_n^2}$ and $\epsilon = \frac{N_c-2}{\gamma+2}\exp\{-\frac{\gamma(\gamma+1)}{\gamma+2}\}$, then:*

$$W(\boldsymbol{\alpha}|\mathbf{x}, \mathbf{J}) - W(\boldsymbol{\alpha}|\mathbf{J}, \mathbf{x}) \geq 1 - e^{-\frac{\|b-s\|^2}{2\sigma_n^2}} - 2\epsilon. \tag{3.94}$$

Proof Following Proposition 3.5, we have:

$$W(\boldsymbol{\beta}|\mathbf{J}, \mathbf{x}) \geq 1 - \frac{1}{2}e^{-\frac{\|b-s\|^2}{2\sigma_n^2}} - \epsilon. \tag{3.95}$$

An upper bound for $W(\boldsymbol{\alpha}|\mathbf{J}, \mathbf{x})$ can be derived as:

$$\begin{aligned} W(\boldsymbol{\alpha}|\mathbf{J}, \mathbf{x}) &= 1 - W(\boldsymbol{\beta}|\mathbf{J}, \mathbf{x}) - \sum_{\hat{\boldsymbol{\alpha}} \neq \boldsymbol{\alpha}, \boldsymbol{\beta}} W(\hat{\boldsymbol{\alpha}}|\mathbf{J}, \mathbf{x}) \\ &\leq 1 - W(\boldsymbol{\beta}|\mathbf{J}, \mathbf{x}) \\ &\leq \frac{1}{2}e^{-\frac{\|b-s\|^2}{2\sigma_n^2}} + \epsilon. \end{aligned} \tag{3.96}$$

It then follows from (3.87) and (3.96) that:

$$W(\boldsymbol{\alpha}|\mathbf{x}, \mathbf{J}) - W(\boldsymbol{\alpha}|\mathbf{J}, \mathbf{x}) \geq 1 - e^{-\frac{\|b-s\|^2}{2\sigma_n^2}} - 2\epsilon. \tag{3.97}$$

■

Proposition 3.6 *Assuming Ω is a PSK constellation with power P_s, let $\mathbf{x} = \boldsymbol{\alpha} s$, $\mathbf{J} = \boldsymbol{\beta} b$, where $\boldsymbol{\alpha}, \boldsymbol{\beta} \in \mathcal{A}$, $\boldsymbol{\alpha} \neq \boldsymbol{\beta}$, and $s, b \in \Omega$, $s \neq b$, then:*

$$W(\boldsymbol{\alpha}|\mathbf{x}, \mathbf{J}) > W(\boldsymbol{\alpha}|\mathbf{J}, \mathbf{x}), \tag{3.98}$$

whenever $\frac{\|b-s\|^2}{\sigma_n^2} > 2\ln\frac{1}{1-2\epsilon}$.

This result follows directly from Theorem 3.2. It implies that as long as s and b are "distinguishable" under the additive noise, the channel symmetricity between the jammer and the legal user is broken, and this increases the probability of correct decision.

Consider $\mathcal{J} = \mathcal{X}$. Define $\hat{W} : \mathcal{X} \times \mathcal{X} \to \mathcal{A}$ by:

$$\hat{W}(\hat{\boldsymbol{\alpha}}|\mathbf{x}, \mathbf{J}) \triangleq \sum_{\mathbf{y}\in\mathcal{Y}} \pi(\mathbf{y}|\mathbf{J}) W(\hat{\boldsymbol{\alpha}}|\mathbf{x}, \mathbf{y}), \tag{3.99}$$

where $\pi : \mathcal{X} \to \mathcal{Y}$ is a probability matrix and $\mathcal{Y} \subseteq \mathcal{X}$. If there exists a π such that:

$$\hat{W}(\hat{\boldsymbol{\alpha}}|\mathbf{x}, \mathbf{J}) = \hat{W}(\hat{\boldsymbol{\alpha}}|\mathbf{J}, \mathbf{x}), \quad \forall \mathbf{x}, \mathbf{J} \in \mathcal{X}, \ \forall \hat{\boldsymbol{\alpha}} \in \mathcal{A}, \tag{3.100}$$

then W is said to be *symmetrizable*. Next, we will show that under ID jamming, as long as the ID sequence is unavailable to the jammer, the AVC corresponding to AJ-MDFH is not only nonsymmetric but also *nonsymmetrizable*.

Note that any probability matrix $\pi : \mathcal{X} \to \mathcal{Y}$ with $\mathcal{Y} \subseteq \mathcal{X}$ can be represented with $\pi : \mathcal{X} \to \mathcal{X}$, as long as we set $\pi(\mathbf{y}|\mathbf{x}) = 0$ for any $\mathbf{x} \in \mathcal{X}, \mathbf{y} \in \mathcal{X} \setminus \mathcal{Y}$. In other words, for any $\mathbf{x}, \mathbf{y} \in \mathcal{X}$, we assume $0 \le \pi(\mathbf{y}|\mathbf{x}) \le 1$. Here the value 1 corresponds to the case that $\mathcal{Y}$ is a single item subset; the value 0 excludes certain points in $\mathcal{X}$ and results in the case that $\mathcal{Y}$ is a proper subset of $\mathcal{X}$. Without loss of generality, in the following, we only consider $\pi : \mathcal{X} \to \mathcal{X}$ under the assumption that $0 \le \pi(\mathbf{y}|\mathbf{x}) \le 1$ for any $\mathbf{x}, \mathbf{y} \in \mathcal{X}$.

Theorem 3.3 *Assuming Ω is an M-PSK constellation with power P_s, let $\gamma = \frac{P_s}{\sigma_n^2}$, $\epsilon = \frac{N_c-2}{\gamma+2}\exp\{-\frac{\gamma(\gamma+1)}{\gamma+2}\}$ and $d_{\min} = \min_{s_1,s_2\in\Omega, s_1\neq s_2} \|s_1 - s_2\|$. Let $f(x) = \frac{1}{x+2}\exp\{-\frac{x(x+1)}{x+2}\}$. For $N_c > 2$ and $M > 2$, under the conditions that:*

$$\gamma > f^{-1}\left(\frac{1}{2N_c}\right) \ \textit{and,}$$

$$\frac{d_{\min}^2}{\sigma_n^2} > \max\left(\frac{2\sqrt{\ln N_c}}{\sqrt{2\gamma} - \sqrt{\ln N_c}}, 2\ln\frac{1}{1-2\epsilon}\right), \tag{3.101}$$

the kernel W for the AVC corresponding to AJ-MDFH is ***nonsymmetrizable***.

We are going to show that for any probability matrix π, there exists some $\hat{\boldsymbol{\alpha}}_0 \in \mathcal{A}$, and $\mathbf{x}_0, \mathbf{J}_0 \in \mathcal{X}$, such that:

$$\hat{W}(\hat{\boldsymbol{\alpha}}_0|\mathbf{x}_0, \mathbf{J}_0) \neq \hat{W}(\hat{\boldsymbol{\alpha}}_0|\mathbf{J}_0, \mathbf{x}_0). \tag{3.102}$$

To prove this result, we need the two lemmas below.

Lemma 3.2 *Assuming $N_c > 2$, $M > 2$, for any given $\pi : \mathcal{X} \to \mathcal{X}$, there exists a pair $\mathbf{x}_0 = \boldsymbol{\alpha} s$ and $\mathbf{J}_0 = \boldsymbol{\beta} b$, $\boldsymbol{\alpha}, \boldsymbol{\beta} \in \mathcal{A}$, $s, b \in \Omega$, such that $\boldsymbol{\beta} \neq \boldsymbol{\alpha}$, $b \neq s$ and $\pi(-\mathbf{x}_0|\mathbf{J}_0) + \pi(\boldsymbol{\beta} s|\mathbf{J}_0) < 1$.*

Proof Suppose for all $\mathbf{x} = \tilde{\boldsymbol{\alpha}}\tilde{s}$ and $\mathbf{J} = \tilde{\boldsymbol{\beta}}\tilde{b}$ with $\tilde{\boldsymbol{\beta}} \neq \tilde{\boldsymbol{\alpha}}, \tilde{b} \neq \tilde{s}$, the equality $\pi(-\mathbf{x}|\mathbf{J}) + \pi(\tilde{\boldsymbol{\beta}}\tilde{s}|\mathbf{J}) = 1$ holds. For $N_c > 2$ and $M > 2$, consider $\mathbf{x}_0 = \boldsymbol{\alpha}s$, $\mathbf{J}_0 = \boldsymbol{\beta}b$ with $\boldsymbol{\beta} \neq \boldsymbol{\alpha}, b \neq s$ and any $\mathbf{x}_1 = \boldsymbol{\lambda}c$, $\boldsymbol{\lambda} \in \mathcal{A}$, $c \in \Omega$ with $\boldsymbol{\lambda} \neq \boldsymbol{\alpha}, \boldsymbol{\beta}$ and $c \neq b, s$. On one hand, $\pi(-\mathbf{x}_0|\mathbf{J}_0) + \pi(\boldsymbol{\beta}s|\mathbf{J}_0) = 1$, which implies that $\mathbf{J}_0$ can only be mapped to $-\mathbf{x}_0$ and $\boldsymbol{\beta}s$. On the other hand, we also have $\pi(-\mathbf{x}_1|\mathbf{J}_0) + \pi(\boldsymbol{\beta}c|\mathbf{J}_0) = 1$, which implies that $\mathbf{J}_0$ can only be mapped to $-\mathbf{x}_1$ and $\boldsymbol{\beta}c$. Since $\mathbf{x}_1 \neq \mathbf{x}_0$ and $\boldsymbol{\beta}c \neq \boldsymbol{\beta}s$, this is a contradiction. Hence, we can always find a pair $\mathbf{x}_0$ and $\mathbf{J}_0$ such that $\pi(-\mathbf{x}_0|\mathbf{J}_0) + \pi(\boldsymbol{\beta}s|\mathbf{J}_0) < 1$. ■

With the same notations as in Lemma 3.2, we have:

Lemma 3.3 *$\mathcal{X}$ can be partitioned into six subsets with respect to $\mathbf{x}_0 = \boldsymbol{\alpha}s$ as $\mathcal{X} = \cup_{i=1}^{6}\mathcal{X}_i$, where*

$$\begin{aligned}
&\mathcal{X}_1 \triangleq \{\boldsymbol{\alpha}(-s)\}, \ \mathcal{X}_2 \triangleq \{\boldsymbol{\alpha}s_0|s_0 \in \Omega, s_0 \neq -s\},\\
&\mathcal{X}_3 \triangleq \{\boldsymbol{\beta}s\}, \ \mathcal{X}_4 \triangleq \{\boldsymbol{\beta}s_0|s_0 \in \Omega, s_0 \neq s\},\\
&\mathcal{X}_5 \triangleq \{\boldsymbol{\alpha}_0 s|\boldsymbol{\alpha}_0 \neq \boldsymbol{\alpha}, \boldsymbol{\beta}\},\\
&\mathcal{X}_6 \triangleq \{\boldsymbol{\alpha}_0 s_0|\boldsymbol{\alpha}_0 \neq \boldsymbol{\alpha}, \boldsymbol{\beta}, s_0 \neq s\}.
\end{aligned} \tag{3.103}$$

Under the conditions that $\gamma > f^{-1}(\frac{1}{2N_c})$ and $\frac{d_{\min}^2}{\sigma_n^2} > \max(\frac{2\sqrt{\ln N_c}}{\sqrt{2}\gamma - \sqrt{\ln N_c}}, 2\ln\frac{1}{1-2\epsilon})$,

$$W(\boldsymbol{\alpha}|\mathbf{x}_0, \mathbf{y}) = W(\boldsymbol{\beta}|\mathbf{x}_0, \mathbf{y}), \ \ \forall \mathbf{y} \in \mathcal{X}_i, i = 1, 3. \tag{3.104}$$

$$W(\boldsymbol{\alpha}|\mathbf{x}_0, \mathbf{y}) - W(\boldsymbol{\beta}|\mathbf{x}_0, \mathbf{y}) > 0, \ \ \forall \mathbf{y} \in \mathcal{X}_i, i = 2, 4, 5, 6. \tag{3.105}$$

Proof See Appendix A. ■

Proof of Theorem 3.3: Following Lemma 3.2, we pick $\mathbf{x}_0$, $\mathbf{J}_0$ such that $\boldsymbol{\beta} \neq \boldsymbol{\alpha}, b \neq s$ and $\pi(-\mathbf{x}_0|\mathbf{J}_0) + \pi(\boldsymbol{\beta}s|\mathbf{J}_0) < 1$. We will prove that $\hat{W}(\boldsymbol{\alpha}|\mathbf{x}_0, \mathbf{J}_0) = \hat{W}(\boldsymbol{\alpha}|\mathbf{J}_0, \mathbf{x}_0)$ and $\hat{W}(\boldsymbol{\beta}|\mathbf{x}_0, \mathbf{J}_0) = \hat{W}(\boldsymbol{\beta}|\mathbf{J}_0, \mathbf{x}_0)$ cannot hold simultaneously, by showing that:

$$\hat{W}(\boldsymbol{\alpha}|\mathbf{x}_0, \mathbf{J}_0) - \hat{W}(\boldsymbol{\beta}|\mathbf{x}_0, \mathbf{J}_0) > \hat{W}(\boldsymbol{\alpha}|\mathbf{J}_0, \mathbf{x}_0) - \hat{W}(\boldsymbol{\beta}|\mathbf{J}_0, \mathbf{x}_0). \tag{3.106}$$

By Lemma 3.3, $\mathcal{X} = \cup_{i=1}^{6}\mathcal{X}_i$. For any $\hat{\boldsymbol{\alpha}}_0 \in \mathcal{A}$, we have:

$$\hat{W}(\hat{\boldsymbol{\alpha}}_0|\mathbf{x}_0, \mathbf{J}_0) = \sum_{i=1}^{6}\sum_{\mathbf{y}\in\mathcal{X}_i} \pi(\mathbf{y}|\mathbf{J}_0)W(\hat{\boldsymbol{\alpha}}_0|\mathbf{x}_0, \mathbf{y}). \tag{3.107}$$

It then follows from (3.104) and (3.105) that:

$$\hat{W}(\boldsymbol{\alpha}|\mathbf{x}_0, \mathbf{J}_0) - \hat{W}(\boldsymbol{\beta}|\mathbf{x}_0, \mathbf{J}_0) = \sum_{\substack{i=2,\\ i\neq 3}}^{6} \sum_{\mathbf{y}\in\mathcal{X}_i} [W(\boldsymbol{\alpha}|\mathbf{x}_0, \mathbf{y}) - W(\boldsymbol{\beta}|\mathbf{x}_0, \mathbf{y})]\pi(\mathbf{y}|\mathbf{J}_0) \geq 0, \tag{3.108}$$

with the equality holds if and only if $\sum_{\substack{i=2,\\ i\neq 3}}^{6} \sum_{\mathbf{y}\in\mathcal{X}_i} \pi(\mathbf{y}|\mathbf{J}_0) = 0$, i.e., $\pi(-\mathbf{x}_0|\mathbf{J}_0) + \pi(\boldsymbol{\beta}s|\mathbf{J}_0) = 1$. Recall that we pick $\mathbf{x}_0$, $\mathbf{J}_0$ such that $\boldsymbol{\beta} \neq \boldsymbol{\alpha}, b \neq s$ and $\pi(-\mathbf{x}_0|\mathbf{J}_0) + \pi(\boldsymbol{\beta}s|\mathbf{J}_0) < 1$. Therefore:

$$\hat{W}(\boldsymbol{\alpha}|\mathbf{x}_0, \mathbf{J}_0) - \hat{W}(\boldsymbol{\beta}|\mathbf{x}_0, \mathbf{J}_0) > 0. \tag{3.109}$$

Similarly, $\mathcal{X}$ can be partitioned into six subsets with respect to $\mathbf{J}_0 = \boldsymbol{\beta}b$, defined as:

$$\begin{aligned}
&\mathcal{J}_1 \triangleq \{\boldsymbol{\beta}(-b)\},\ \mathcal{J}_2 \triangleq \{\boldsymbol{\beta}b_0|b_0 \in \Omega, b_0 \neq -b\},\\
&\mathcal{J}_3 \triangleq \{\boldsymbol{\alpha}b\},\ \mathcal{J}_4 \triangleq \{\boldsymbol{\alpha}b_0|b_0 \in \Omega, b_0 \neq b\},\\
&\mathcal{J}_5 \triangleq \{\boldsymbol{\beta}_0 b|\boldsymbol{\beta}_0 \neq \boldsymbol{\alpha}, \boldsymbol{\beta}\},\ \mathcal{J}_6 \triangleq \{\boldsymbol{\beta}_0 b_0|\boldsymbol{\beta}_0 \neq \boldsymbol{\alpha}, \boldsymbol{\beta}, b_0 \neq b\},
\end{aligned} \tag{3.110}$$

thus:

$$\hat{W}(\hat{\boldsymbol{\alpha}}_0|\mathbf{J}_0, \mathbf{x}_0) = \sum_{i=1}^{6} \sum_{\mathbf{y}\in\mathcal{J}_i} \pi(\mathbf{y}|\mathbf{x}_0) W(\hat{\boldsymbol{\alpha}}_0|\mathbf{J}_0, \mathbf{y}). \tag{3.111}$$

Then we have:

$$\begin{aligned}
&\hat{W}(\boldsymbol{\alpha}|\mathbf{J}_0, \mathbf{x}_0) - \hat{W}(\boldsymbol{\beta}|\mathbf{J}_0, \mathbf{x}_0)\\
&= \sum_{\substack{i=2,\\ i\neq 3}}^{6} \sum_{\mathbf{y}\in\mathcal{J}_i} [W(\boldsymbol{\alpha}|\mathbf{J}_0, \mathbf{y}) - W(\boldsymbol{\beta}|\mathbf{J}_0, \mathbf{y})]\pi(\mathbf{y}|\mathbf{x}_0).
\end{aligned} \tag{3.112}$$

Moreover, under the same conditions as in previous case:

$$\hat{W}(\boldsymbol{\alpha}|\mathbf{J}_0, \mathbf{x}_0) - \hat{W}(\boldsymbol{\beta}|\mathbf{J}_0, \mathbf{x}_0) \leq 0. \tag{3.113}$$

Therefore, we can see that (3.106) holds, which implies that $\hat{W}(\boldsymbol{\alpha}|\mathbf{x}_0, \mathbf{J}_0) = \hat{W}(\boldsymbol{\alpha}|\mathbf{J}_0, \mathbf{x}_0)$ and $\hat{W}(\boldsymbol{\beta}|\mathbf{x}_0, \mathbf{J}_0) = \hat{W}(\boldsymbol{\beta}|\mathbf{J}_0, \mathbf{x}_0)$ cannot hold simultaneously. ■

Note that the secure ID in AJ-MDFH is generated using AES, to symmetrize AJ-MDFH is thus equivalent to breaking AES, which is computationally infeasible in practical systems. That is, the AVC corresponding to AJ-MDFH is computationally infeasible to be symmetrized. This result ensures that when the ID sequence is unknown to the jammer, the deterministic capacity of AJ-MDFH is positive and equal to the random code capacity [37, 39].

3.5.2.2 Capacity Calculation

Note that in AJ-MDFH, the message information is only transmitted through the carrier bits. Consider $\mathbf{x} = \boldsymbol{\alpha} s$ where $s \in \Omega$ and $\boldsymbol{\alpha} = (\alpha_1, \cdots, \alpha_{N_c}) \in \mathcal{A}$. Let i_S and i_J be the signal channel index and jamming channel index, respectively, and $\hat{i}_S$ the detected signal channel index at the receiver. For capacity derivation, define

$$W_1(\hat{k}|k, j) \triangleq Pr\{\hat{i}_S = \hat{k}|i_S = k, i_J = j\}. \tag{3.114}$$

Let $\mathbf{x} = \boldsymbol{\alpha} s$, $\mathbf{J} = \boldsymbol{\beta} b$ with $\boldsymbol{\alpha} = v(k)$, $\boldsymbol{\beta} = v(j)$. Let $\hat{\boldsymbol{\alpha}} = v(\hat{k})$, and assuming s and b are uniformly distributed over Ω, then the relationship between W_1 and W can be characterized as:

$$W_1(\hat{k}|k, j) = \frac{1}{|\Omega|^2} \sum_{s\in\Omega} \sum_{b\in\Omega} W(\hat{\boldsymbol{\alpha}}|\mathbf{x} = \boldsymbol{\alpha} s, \mathbf{J} = \boldsymbol{\beta} b). \tag{3.115}$$

The detailed representation of W_1 is provided in Appendix B, where we prove that W_1 has the following properties:

(P1): $W_1(k|k, k) = W_1(k_0|k_0, k_0)$ and $W_1(i|k, k) = W_1(i_0|k_0, k_0)$ for any $i, k, i_0, k_0 \in \mathcal{I}_c, i \neq k, i_0 \neq k_0$.

(P2): $W_1(k|k, j) = W_1(k_0|k_0, j_0)$, $W_1(j|k, j) = W_1(j_0|k_0, j_0)$ and $W_1(i|k, j) = W_1(i_0|k_0, j_0)$ for any $i, j, k, i_0, j_0, k_0 \in \mathcal{I}_c$, $j \neq k, i \neq j, k$, $j_0 \neq k_0, i_0 \neq j_0, k_0$.

Denote the set of all probability distributions on $\mathcal{I}_c$ as $\mathcal{P}(\mathcal{I}_c)$. Let P and ζ denote the probability distribution associated with i_S and i_J, respectively. $P, \zeta \in \mathcal{P}(\mathcal{I}_c)$. Let W_ζ denote the averaged probability matrix for a given ζ

$$\begin{aligned} W_\zeta(\hat{k}|k) &= W_\zeta(\hat{i}_S = \hat{k}|i_S = k) \\ &= \sum_{j\in\mathcal{I}_c} W_1(\hat{k}|k, j)\zeta(i_J = j). \end{aligned} \tag{3.116}$$

Let $I(P, W_\zeta)$ denote the mutual information [37] between the input and the output for the AJ-MDFH channel, defined as:

$$I(P, W_\zeta) \triangleq \sum_{\hat{k}\in\mathcal{I}_c} \sum_{k\in\mathcal{I}_c} P(i_S = k) W_\zeta(\hat{k}|k) \log \frac{W_\zeta(\hat{k}|k)}{(PW)_\zeta(\hat{k})}, \tag{3.117}$$

where $(PW)_\zeta(\hat{k}) = \sum_{k' \in \mathcal{I}_c} W_\zeta(\hat{k}|k')P(k')$. Following Theorem 3.3, the AVC corresponding to AJ-MDFH is nonsymmetrizable. Its channel capacity for the average error probability is positive and can be calculated as: [36, 40]

$$C = \max_{P \in \mathcal{P}(\mathcal{I}_c)} \min_{\zeta \in \mathcal{P}(\mathcal{I}_c)} I(P, W_\zeta) = \min_{\zeta \in \mathcal{P}(\mathcal{I}_c)} \max_{P \in \mathcal{P}(\mathcal{I}_c)} I(P, W_\zeta). \tag{3.118}$$

It can be observed from (3.118) that the legal user tries to choose P to maximize the mutual information, while the jammer tries to minimize it by choosing an appropriate ζ. Let $(P, \zeta) \in \mathcal{P}(\mathcal{I}_c) \times \mathcal{P}(\mathcal{I}_c)$ be a pair of mixed strategy chosen by the user and the jammer. The capacity can be achieved when a pair of saddle point strategy (P^*, ζ^*) are chosen, which can be characterized by the following two inequalities for all $(P, \zeta) \in \mathcal{P}(\mathcal{I}_c) \times \mathcal{P}(\mathcal{I}_c)$ [41–44]:

$$I(P, W_{\zeta^*}) \leq I(P^*, W_{\zeta^*}) \leq I(P^*, W_\zeta). \tag{3.119}$$

Following the same argument as in [45], it can be shown that:

Lemma 3.4 *In an AJ-MDFH channel, the saddle point strategy pair can be reached when both P and ζ are uniform distributions over $\mathcal{I}_c$. That is:*

$$P^*(k) = \begin{cases} \frac{1}{N_c}, & k \in \mathcal{I}_c, \\ 0, & \text{otherwise}, \end{cases} \qquad \zeta^*(j) = \begin{cases} \frac{1}{N_c}, & j \in \mathcal{I}_c, \\ 0, & \text{otherwise}. \end{cases} \tag{3.120}$$

In AJ-MDFH, when the jammer chooses the strategy ζ^* as in (3.120), the averaged probability matrix can be calculated as:

$$W_{\zeta^*}(\hat{k}|k) = \sum_{j=1}^{N_c} W_1(\hat{k}|k, j)\zeta^*(j). \tag{3.121}$$

(i) When $\hat{k} = k$, (3.121) can be expanded as:

$$W_{\zeta^*}(\hat{k}|k) = W_1(k|k, k)\zeta^*(k) + \sum_{j \in \mathcal{I}_c, j \neq k} W_1(k|k, j)\zeta^*(j). \tag{3.122}$$

Following the properties **(P1)** and **(P2)** of W_1, we have $W_{\zeta^*}(\hat{k}|k) = \frac{1}{N_c} W_1(k_0|k_0, k_0) + \frac{N_c - 1}{N_c} W_1(k_0|k_0, j_0)$, for any fixed $j_0, k_0 \in \mathcal{I}_c$, $j_0 \neq k_0$.

(ii) When $\hat{k} \neq k$, (3.121) can be expanded as:

$$W_{\zeta^*}(\hat{k}|k) = W_1(\hat{k}|k, k)\zeta^*(k) + W_1(\hat{k}|k, \hat{k})\zeta^*(\hat{k}) + \sum_{j \in \mathcal{I}_c, j \neq \hat{k}, k} W_1(\hat{k}|k, j)\zeta^*(j). \tag{3.123}$$

Following the properties **(P1)** and **(P2)** of W_1, we have $W_{\zeta^*}(\hat{k}|k) = \frac{1}{N_c}W_1(\hat{k}_0|k_0, k_0) + \frac{1}{N_c}W_1(\hat{k}_0|k_0, \hat{k}_0) + \frac{N_c-2}{N_c}W_1(\hat{k}_0|k_0, j_0)$, for any fixed $\hat{k}_0, k_0, j_0 \in \mathcal{I}_c, \hat{k}_0 \neq k_0, j_0 \neq \hat{k}_0, k_0$. Define $w_1 \triangleq W_{\zeta^*}(k|k)$ and $w_2 \triangleq W_{\zeta^*}(\hat{k}|k), \hat{k} \neq k$, and then W_{ζ^*} can be obtained as:

$$W_{\zeta^*} = \begin{pmatrix} W_{\zeta^*}(1|1) & W_{\zeta^*}(2|1) & \cdots & W_{\zeta^*}(N_c|1) \\ W_{\zeta^*}(1|2) & W_{\zeta^*}(2|2) & \cdots & W_{\zeta^*}(N_c|2) \\ \vdots & \vdots & \ddots & \vdots \\ W_{\zeta^*}(1|N_c) & W_{\zeta^*}(2|N_c) & \cdots & W_{\zeta^*}(N_c|N_c) \end{pmatrix} = \begin{pmatrix} w_1 & w_2 & \cdots & w_2 \\ w_2 & w_1 & \cdots & w_2 \\ \vdots & \vdots & \ddots & \vdots \\ w_2 & w_2 & \cdots & w_1 \end{pmatrix}_{N_c \times N_c}. \tag{3.124}$$

Due to the special structure of matrix W_{ζ^*}, we have: for any $\hat{k}, k' \in \mathcal{I}_c$, $\sum_{k' \in \mathcal{I}_c} W_{\zeta^*}(\hat{k}|k') = \sum_{\hat{k} \in \mathcal{I}_c} W_{\zeta^*}(\hat{k}|k') = 1$, and

$$(P^*W)_{\zeta^*}(\hat{k}) = \sum_{k' \in \mathcal{I}_c} W_{\zeta^*}(\hat{k}|k')P^*(k') = \frac{1}{N_c}. \tag{3.125}$$

Therefore, the capacity can be calculated as:

$$\begin{aligned} C &= I(P^*, W_{\zeta^*}) \\ &= \sum_{\hat{k} \in \mathcal{I}_c} \sum_{k \in \mathcal{I}_c} \frac{1}{N_c} W_{\zeta^*}(\hat{k}|k) \log \frac{W_{\zeta^*}(\hat{k}|k)}{\frac{1}{N_c}} \\ &= \sum_{\hat{k} \in \mathcal{I}_c} \sum_{k \in \mathcal{I}_c} \frac{1}{N_c} W_{\zeta^*}(\hat{k}|k) \log N_c + \sum_{\hat{k} \in \mathcal{I}_c} \sum_{k \in \mathcal{I}_c} \frac{1}{N_c} W_{\zeta^*}(\hat{k}|k) \log W_{\zeta^*}(\hat{k}|k) \\ &= \log N_c + \sum_{\hat{k}=1}^{N_c} W_{\zeta^*}(\hat{k}|1) \log W_{\zeta^*}(\hat{k}|1) \\ &= \log N_c + w_1 \log w_1 + (N_c - 1) w_2 \log w_2. \end{aligned} \tag{3.126}$$

Following the discussions above, we have

Theorem 3.4 *Assuming Ω is an M-PSK constellation with power P_s, under the worst-case single band disguised jamming, the channel capacity of AJ-MDFH*

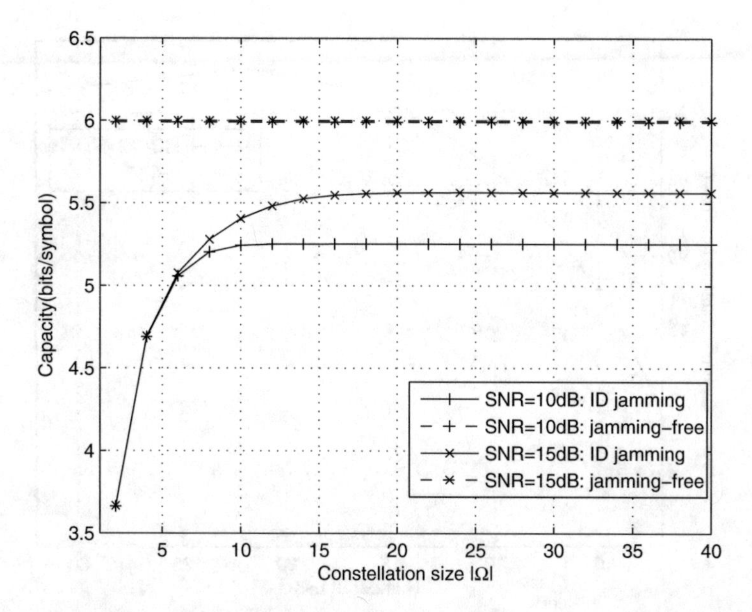

Fig. 3.13 AJ-MDFH capacity under the worst-case single band disguised jamming (ID jamming) for different PSK constellation size. $N_c = 64$. (© [2013] IEEE. Reprinted, with permission, from Ref. [22])

system is a function of M, N_c *and* $\frac{P_s}{\sigma_n^2}$ *of the form* $C = C\left(M, N_c, \frac{P_s}{\sigma_n^2}\right)$. *As* M *approaches infinity,* C *converges to*

$$\bar{C} = \log N_c + \bar{w}_1 \log \bar{w}_1 + (N_c - 1)\bar{w}_2 \log \bar{w}_2, \tag{3.127}$$

where $\bar{w}_1 = \lim_{M\to\infty} w_1$ *and* $\bar{w}_2 = \lim_{M\to\infty} w_2$.

The convergence result follows from similar argument as in the proof of Theorem 3.1. Our analysis in this section can be extended to MC-AJ-MDFH, which is a secure combination of several collision-free single carrier AJ-MDFH systems. The capacity of MC-AJ-MDFH can be obtained as:

$$C_{MC} = \sum_{m=1}^{N_g} C_m, \tag{3.128}$$

where N_g is the number of carriers and C_m is the capacity of the m-th carrier.

Theorem 3.4 is illustrated in Fig. 3.13, where we can see that under reasonable SNR levels (e.g., $\geq$10 dB), the capacity limit $\bar{C}$ is close to the corresponding jamming-free case indicated by the dashed line. Figure 3.14 compares the capacity of MC-AJ-MDFH and frequency hopping multiple access (FHMA) system in

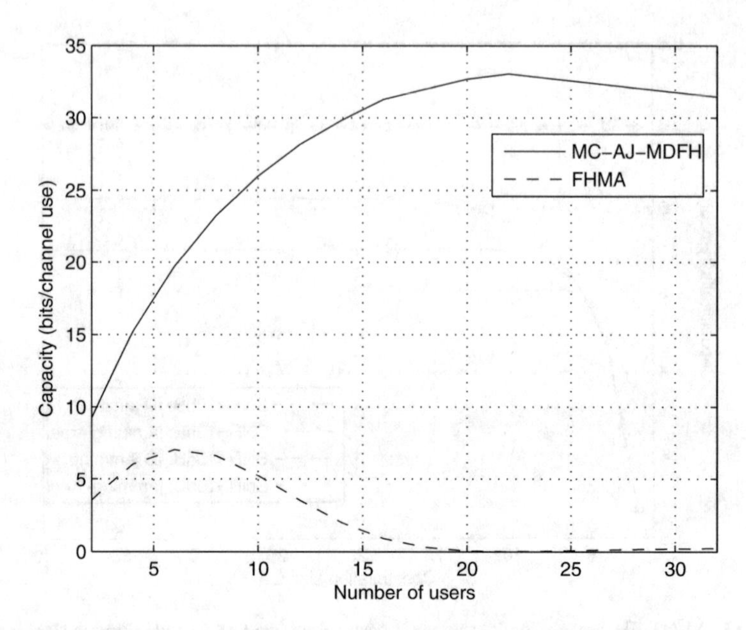

Fig. 3.14 Capacity of MC-AJ-MDFH and FHMA under the worst-case single band disguised jamming. $N_c = 64$, $SNR = 10\,\text{dB}$. Here, per channel use means the total bandwidth of all used channels over one hopping period. (© [2013] IEEE. Reprinted, with permission, from Ref. [22])

[2, 46]. It can be observed that due to the collision-free design and the use of ID sequence, under disguised jamming, MC-AJ-MDFH can effectively support much more users than FHMA.

3.6 Conclusions

In this chapter, first, we introduced a three-dimensional modulation scheme, message-driven frequency hopping (MDFH). The basic idea is that the selection of carrier frequencies is directly controlled by the encrypted information stream rather than by a pre-selected pseudo-random sequence as in conventional FH. It was shown that transmission through hopping frequency control adds another dimension to the signal space, and the resulted coding gain can increase the system spectral efficiency by multiple times.

Second, we analyzed the performance of MDFH under hostile jamming. It was observed that MDFH is particularly powerful under strong jamming scenarios and outperforms the conventional FH by big margins. However, when the system experiences *disguised jamming*, where the jamming is highly correlated with the signal, and has a power level close or equal to the signal power, it is then difficult for the MDFH receiver to distinguish jamming from the true signal, resulting in significant performance losses.

Third, to enhance the jamming resistence of MDFH, we introduced the anti-jamming MDFH (AJ-MDFH) system. We showed that by inserting a secure ID sequence in transmission, AJ-MDFH can effectively reduce the performance degradation caused by disguised jamming. The impact of ID constellation design on system performance was investigated under both noise jamming and ID jamming. It was proved that for a given power constraint, constant modulus constellation delivers the best results under noise jamming in terms of detection error probability; while under ID jamming, when noise is present, the detection error probability converges as the constellation size goes to infinity. Moreover, AJ-MDFH can be extended to MC-AJ-MDFH by allowing simultaneous multi-carrier transmission. With jamming randomization and enriched frequency diversity, MC-AJ-MDFH can increase the system efficiency and jamming resistance significantly and can readily be used as a collision-free multiple access scheme.

Finally, we analyzed the capacity of MDFH and AJ-MDFH under disguised jamming. We proved that under the worst-case disguised jamming, (i) for MDFH, the corresponding AVC is symmetric, which implies that the deterministic capacity of MDFH is zero; and (ii) for AJ-MDFH, due to shared randomness between the transmitter and the receiver provided by the secure ID sequence, the corresponding AVC is nonsymmetrizable, which implies that the deterministic capacity of AJ-MDFH is positive and equal to the random code capacity. We calculated the capacity of AJ-MDFH and showed that it converges as the ID constellation size goes to infinity. This echoes our previous result that the probability of error of AJ-MDFH converges as the ID constellation size goes to infinity. In this chapter, again, we can see that shared secure randomness between the transmitter and the receiver plays a critical role in anti-jamming system design.

References

1. G.R. Cooper and R.W. Nettleton. A spread spectrum technique for high capacity mobile communuzation. *IEEE Transactions on Vehicular Technology*, 27(4):264–275, Nov. 1978.
2. A.J. Viterbi. A processing-satellite transponder for multlple access by low rate mobile users. In *Proc. Digital Satellite Commun. Conf.*, pages 166–174, Montreal, Canada, October 1978.
3. E.A. Geraniotis and M.B. Pursley. Error probabilities for slow frequency-hopped spread-spectrum multiple-access communication over fading channels. *IEEE Transactions on Communications*, 30(5):996–1009, May 1982.
4. E.A. Geraniotis. Multiple-access capability of frequency-hopped spread-spectrum revisited: An analysis of the effect of unequal power levels. *IEEE Transactions on Communications*, 38(7):1066–1077, Jul. 1990.
5. M.A. Wickert and R.L. Turcotte. Probability of error analysis for FHSS/CDMA communications in the presence of fading. *IEEE Journal on Selected Areas in Communications*, 10(3):523–534, Apr. 1992.
6. Y.R. Tsai and J.F. Chang. Using frequency hopping spread spectrum technique to combat multipath interference in a multiaccessing environment. *IEEE Transactions on Vehicular Technology*, 43(2):211–222, May 1994.
7. F. Dominique and J.H. Reed. Robust frequency hop synchronisation algorithm. *Electronics Letters*, 32:1450–1451, August 1996.

8. M. Simon, G. Huth, and A. Polydoros. Differentially coherent detection of QASK for frequency-hopping systems–part i: Performance in the presence of a Gaussian noise environment. *IEEE Transactions on Communications*, 30:158–164, January 1982.
9. Y.M. Lam and P.H. Wittke. Frequency-hopped spread-spectrum transmission with band-efficient modulations and simplified noncoherent sequence estimation. *IEEE Transactions on Communications*, 38:2184–2196, December 1990.
10. Joonyoung Cho, Youhan Kim, and Kyungwhoon Cheun. A novel FHSS multiple-access network using M-ary orthogonal Walsh modulation. In *Proc. 52nd IEEE Veh. Technol. Conf.*, volume 3, pages 1134–1141, Sept. 2000.
11. S. Glisic, Z. Nikolic, N. Milosevic, and A. Pouttu. Advanced frequency hopping modulation for spread spectrum WLAN. *IEEE Journal on Selected Areas in Communications*, 18:16–29, January 2000.
12. Kwonhue Choi and Kyungwhoon Cheun. Maximum throughput of FHSS multiple-access networks using MFSK modulation. *IEEE Transactions on Communications*, 52(3):426–434, March 2004.
13. Kang-Chun Peng, Chien-Hsiang Huang, Chien-Jung Li, and Tzyy-Sheng Horng. High-performance frequency-hopping transmitters using two-point Delta-Sigma modulation. *IEEE Transactions on Microwave Theory and Techniques*, 52:2529–2535, November 2004.
14. K. Choi and K. Cheun. Optimum parameters for maximum throughput of FHMA system with multilevel FSK. *IEEE Transactions on Vehicular Technology*, 55:1485–1492, Sept. 2006.
15. Qi Ling and Tongtong Li. Message-driven frequency hopping: Design and analysis. *IEEE Transactions on Wireless Communications*, 8(4):1773–1782, April 2009.
16. F.H.P. Fitzek. The medium is the message. In *Proc. IEEE Intl. Conf. Commun.*, volume 11, pages 5016–5021, June 2006.
17. Tongtong Li, Qi Ling, and Jian Ren. Physical layer built-in security analysis and enhancement algorithms for CDMA systems. *EURASIP Journal on Wireless Communications and Networking*, 2007:Article ID 83589, 7 pages, 2007.
18. John G. Proakis. *Digital Communications*. McGraw-Hill, fourth edition, 2000.
19. D.I. Goodman, P.S. Henry, and V.K. Prabhu. Frequency-hopped multilevel FSK for mobile radio. *Bell System Technical Journal*, 59:1257–1275, Sept. 1980.
20. J.I. Marcum. Table of Q functions. *U.S. Air Force Project RAND Res. Memo. M-339*, January 1950. ASTIA Document AD 1165451, Rand Corp.
21. Lei Zhang, Huahui Wang, and Tongtong Li. Anti-jamming message-driven frequency hopping: part I – system design. *IEEE Transactions on Wireless Communications*, pages 70–79, 2013.
22. Lei Zhang and Tongtong Li. Anti-jamming message-driven frequency hopping: part II — capacity analysis under disguised jamming. *IEEE Transactions on Wireless Communications*, Vol. 12, No. 1, pages 80–88, 2013.
23. Advanced encryption standard. ser. FIPS-197, November 2001.
24. Leonard Lightfoot, Lei Zhang, and Tongtong Li. Secure collision-free frequency hopping for OFDMA based wireless networks. *EURASIP Journal on Advances in Signal Processing*, 2009, 2009.
25. R. Viswanathan and K. Taghizadeh. Diversity combining in FH/BFSK systems to combat partial band jamming. *IEEE Transactions on Communications*, 36(9):1062–1069, Sep 1988.
26. Jhong Lee, L. Miller, and Young Kim. Probability of error analyses of a BFSK frequency-hopping system with diversity under partial-band jamming interference–part II: Performance of square-law nonlinear combining soft decision receivers. *IEEE Transactions on Communications*, 32(12):1243–1250, Dec 1984.
27. L. Miller, Jhong Lee, and A. Kadrichu. Probability of error analyses of a BFSK frequency-hopping system with diversity under partial-band jamming interference–part III: Performance of a square-law self-normalizing soft decision receiver. *IEEE Transactions on Communications*, 34(7):669–675, Jul 1986.
28. Jhong Lee and L. Miller. Error performance analyses of differential phase-shift-keyed/frequency-hopping spread-spectrum communication system in the partial-band jamming environments. *IEEE Transactions on Communications*, 30(5):943–952, May 1982.

29. J.J. Kang and K.C. Teh. Performance of coherent fast frequency-hopped spread-spectrum receivers with partial-band noise jamming and AWGN. *IEE Proceedings–Communications*, 152(5):679–685, Oct. 2005.
30. C. Esli and H. Delic. Antijamming performance of space-frequency coding in partial-band noise. *IEEE Transactions on Vehicular Technology*, 55(2):466–476, March 2006.
31. S. Stein. Unified analysis of certain coherent and noncoherent binary communications systems. *IEEE Transactions on Information Theory*, 10(1):43–51, Jan 1964.
32. Marek Kuczma. *An Introduction to the Theory of Functional Equations and Inequalities: Cauchy's Equation and Jensen's Inequality*. Springer, second edition, 2009.
33. D. Blackwell, L. Breiman, and AJ Thomasian. The capacities of certain channel classes under random coding. *The Annals of Mathematical Statistics*, pages 558–567, 1960.
34. I. Csiszar and J. Körner. *Information theory: coding theorems for discrete memoryless systems*, volume 244. Academic press, 1981.
35. I. Csiszár and P. Narayan. Arbitrarily varying channels with constrained inputs and states. *IEEE Transactions on Information Theory*, 34(1):27–34, 1988.
36. I. Csiszar and P. Narayan. The capacity of the arbitrarily varying channel revisited: Positivity, constraints. *IEEE Transactions on Information Theory*, 34(2):181–193, 1988.
37. A. Lapidoth and P. Narayan. Reliable communication under channel uncertainty. *IEEE Transactions on Information Theory*, 44(6):2148–2177, Oct. 1998.
38. A.D. Sarwate. Robust and adaptive communication under uncertain interference. Technical report, Technical Report No. UCB/EECS-2008-86, University of California at Berkeley, 2008.
39. T. Ericson. Exponential error bounds for random codes in the arbitrarily varying channel. *IEEE Transactions on Information Theory*, 31(1):42–48, Jan. 1985.
40. R. Ahlswede. Elimination of correlation in random codes for arbitrarily varying channels. *Probability Theory and Related Fields*, 44(2):159–175, 1978.
41. J.M. Borden, D.M. Mason, and R.J. McEliece. Some information theoretic saddlepoints. *SIAM journal on control and optimization*, 23:129, 1985.
42. T Basar and Y W Wu. Solutions to a class of minimax decision problems arising in communications systems. *J. Optim. Theory Appl.*, 51:375–404, Decr 1986.
43. T. Başar. The Gaussian test channel with an intelligent jammer. *IEEE Transactions on Information Theory*, 29(1):152–157, 1983.
44. T. Başar and G.J. Olsder. *Dynamic noncooperative game theory*, volume 23. Society for Industrial Mathematics, 1999.
45. I. Stiglitz. Coding for a class of unknown channels. 12(2):189–195, apr 1966.
46. J.G. Goh and S.V. Maric. The capacities of frequency-hopped code-division multiple-access channels. *IEEE Transactions on Information Theory*, 44(3):1204–1211, 1998.

Chapter 4
Collision-Free Frequency Hopping and OFDM

4.1 Enhance Jamming Resistance: The Combination of OFDM and Frequency Hopping

Orthogonal frequency-division multiplexing (OFDM) and the associated orthogonal frequency-division multiple-access (OFDMA) scheme have become the major air interfaces for broadband wireless communications [1]. By dividing the entire channel into mutually orthogonal parallel subchannels, the OFDM technique transforms a frequency-selective fading channel into parallel flat-fading channels. As a result, OFDM can effectively eliminate the intersymbol interference (ISI) caused by the multipath environment with simple transceiver design and, at the same time, achieve high spectral efficiency. For this reason, OFDM is used in the wireless local area network (WLAN) technologies based on IEEE 802.11 standard series, the wireless metropolitan area network (WMAN) technologies based on the IEEE 802.16 standard series, and the long-term evolution (LTE) cellular broadband technology [1]. However, OFDMA does not possess any inherent security features and is fragile to hostile jamming.

On the other hand, the frequency hopping (FH) is originally designed for jamming resistant communications. In traditional FH systems, the transmitter hops in a pseudorandom manner among available frequencies according to a pre-specified algorithm, and the receiver then operates in a strict synchronization with the transmitter and remains tuned to the same center frequency. Two major limitations with the conventional FH scheme are (i) *strong requirement on frequency acquisition.* In existing FH systems, exact frequency synchronization has to be kept between the transmitter and the receiver. The strict requirement on synchronization directly influences the complexity, design, and performance of the system [2] and turns out to be a significant challenge in fast hopping system design. (ii) *Low spectral efficiency over large bandwidth.* Typically, FH systems require large bandwidth, which is proportional to the hopping rate and the number of all the available channels. In conventional frequency hopping multiple access (FHMA), each user

T. Li et al., *Wireless Communications Under Hostile Jamming: Security and Efficiency*, https://doi.org/10.1007/978-981-13-0821-5_4

hops independently based on its own pseudorandom number (PN) sequence, and a collision occurs whenever there are two users over the same frequency band. Mainly limited by the collision effect, the spectral efficiency of conventional FH systems is very low. In literature, considerable efforts have been devoted to increasing the spectral efficiency of FH systems by applying high-dimensional modulation schemes [3–9].

More recently, a combination of the FH technique and the OFDMA system, called FH-OFDMA, has been proposed [10, 11]. This system is based on the conventional FH techniques, and the spectral efficiency is seriously limited by the collision effect. Along with the ever-increasing demand on inherently secure high data-rate wireless communications, new techniques that are more efficient and reliable have to be developed.

In this chapter, we consider OFDM-based highly efficient anti-jamming system design using FH-based secure dynamic spectrum access control. Both partial-band jamming and full-band jamming are considered.

First, we present a collision-free frequency hopping (CFFH) system based on the OFDMA framework and an innovative secure subcarrier assignment scheme. The secure subcarrier assignment is achieved through an Advanced Encryption Standard (AES) [12]-based secure permutation algorithm, which is designed to ensure that (i) each user hops to a new set of subcarriers in a pseudorandom manner at the beginning of each hopping period, (ii) different users always transmit on non-overlapping sets of subcarriers, and (iii) malicious users cannot determine the hopping pattern of the authorized users and hence cannot launch follower jamming attacks.[1] In other words, the CFFH scheme can effectively mitigate jamming interference, including both random jamming and follower jamming. Moreover, using the fast Fourier transform (FFT)-based OFDMA framework, CFFH has the same high spectral efficiency as that of OFDM and at the same time can relax the complex frequency synchronization problem suffered by conventional FH systems.

We further enhance the anti-jamming property of CFFH by incorporating the space-time coding (STC) scheme. Space-time block coding, which was first proposed by Alamouti [13] and refined by Tarokh et al. [14, 15], is a technique that exploits antenna array spatial diversity to provide gains against fading environments. When incorporated with OFDM, the space-time diversity in space-time coding is then converted to space-frequency diversity. The combination of space-time coding and CFFH is found to be particularly powerful in eliminating channel interference and hostile jamming interference, especially random jamming. In this chapter, we analyze the performance of the STC-CFFH system through the following aspects: (i) comparing the spectral efficiency of STC-CFFH with that of the conventional FH-

[1] Follower jamming is the worst jamming scenario, in which the attacker is aware of the carrier frequency or the frequency hopping pattern of an authorized user and can destroy the user's communication by launching jamming interference over the same frequency bands.

OFDMA system and (ii) investigating the performance of the STC-CFFH system under Rayleigh fading with hostile jamming. Our analysis indicates that the STC-CFFH system is both highly efficient and very robust under jamming environments.

4.2 Secure Subcarrier Assignment

In this section, we present the secure subcarrier assignment scheme, for which the major component is an AES-based secure permutation algorithm. AES is chosen because of its simplicity of design, variable block and key sizes, feasibility in both hardware and software, and resistance against all known attacks [16]. Note that, the secure subcarrier assignment is not limited to any particular cryptographic algorithm but is highly recommended that only thoroughly analyzed cryptographic algorithms be applied.

The AES-based permutation algorithm is used to securely select the frequency hopping pattern for each user so that (i) different users always transmit on non-overlapping sets of subcarriers and (ii) malicious users cannot determine the frequency hopping pattern and therefore cannot launch follower jamming attacks.

We assume there are a total of N_c available subcarriers and there are M users in the system. For $i = 0, 1, \cdots, M-1$, the number of subcarriers assigned to user i is denoted as N_u^i. We assume that different users transmit over non-overlapping set of subcarriers, and we have $\sum_{i=0}^{M-1} N_u^i = N_c$. The secure subcarrier assignment algorithm is described in the following subsections.

4.2.1 Secure Permutation Index Generation

A pseudorandom binary sequence is generated using a 32-bit linear feedback shift register (LFSR), which is initialized by a secret sequence chosen by the base station. The LFSR has the following characteristic polynomial:

$$\begin{aligned} &x^{32} + x^{26} + x^{23} + x^{22} + x^{16} + x^{12} + x^{11} \\ &+x^{10} + x^8 + x^7 + x^5 + x^4 + x^2 + x + 1. \end{aligned} \tag{4.1}$$

Use the pseudorandom binary sequence generated by the LFSR as the plaintext. Encrypt the plaintext using the AES algorithm and a secure key. The key size can be 128, 192, or 256. The encrypted plaintext is known as the ciphertext. Assume N_c is a power of 2; pick an integer $L \in [\frac{N_c}{2}, N_c]$. Note that a total of $N_b = \log_2 N_c$ bits are required to represent each subcarrier; let $q = L\log_2 N_c$. Take q bits from the ciphertext, and put them as a q-bit vector $\mathbf{e} = [e_1, e_2, \cdots, e_q]$.

Partition the ciphertext sequence $\mathbf{e}$ into L groups, such that each group contains N_b bits. For $k = 1, 2, \cdots, L$, the partition of the ciphertext is as follows:

$$\mathbf{p_k} = [e_{(k-1)*N_b+1}, e_{(k-1)*N_b+2}, \cdots, e_{(k-1)*N_b+N_b}], \tag{4.2}$$

where $\mathbf{p_k}$ corresponds to the kth N_b-bit vector.

For $k = 1, 2, \cdots, L$ denote P_k as the decimal number corresponding to $\mathbf{p_k}$, such that

$$\begin{aligned} P_k = & \, e_{(k-1)*N_b+1} \cdot 2^{N_b-1} + e_{(k-1)*N_b+2} \cdot 2^{N_b-2} \\ & + \cdots + e_{(k-1)*N_b+N_b-1} \cdot 2^1 \\ & + e_{(k-1)*N_b+N_b} \cdot 2^0 . \end{aligned} \tag{4.3}$$

Finally, we denote $P = [P_1, P_2, \cdots, P_L]$ as the permutation index vector. Here the largest number in P is $N_c - 1$. In the following subsection, we will discuss the secure permutation algorithm.

4.2.2 Secure Permutation Algorithm and Subcarrier Assignment

For $k = 0, 1, 2, \cdots, L$, denote $I_k = [I_k(0), I_k(1), \cdots, I_k(N_c - 1)]$ as the index vector at the kth step. The secure permutation scheme of the index vector is achieved through the following steps:

0. Initially, the index vector is $I_0 = [I_0(0), I_0(1), \cdots, I_0(N_c - 1)]$, and the permutation index is $P = [P_1, P_2, \cdots, P_L]$. We start with $I_0 = [0, 1, \cdots, N_c - 1]$.
1. For $k = 1$, switch $I_0(0)$ and $I_0(P_1)$ in index vector I_0 to obtain I_1. In other words, $I_1 = [I_1(0), I_1(1), \cdots, I_1(N_c - 1)]$, where $I_1(0) = I_0(P_1)$, $I_1(P_1) = I_0(0)$ and $I_1(m) = I_0(m)$ for $m \neq 0, P_1$.
2. Repeat the previous step for $k = 2, 3, \cdots, L$. In general, if we already have $I_{k-1} = [I_{k-1}(0), I_{k-1}(1), \cdots, I_{k-1}(N_c - 1)]$, then we can obtain $I_k = [I_k(0), I_k(1), \cdots, I_k(N_c - 1)]$ through the permutation defined as $I_k(k-1) = I_{k-1}(P_k)$, $I_k(P_k) = I_{k-1}(k-1)$ and $I_k(m) = I_{k-1}(m)$ for $m \neq k-1, P_k$.
3. After L steps, we obtain the subcarrier frequency vector as $F_L = [f_{I_L(0)}, f_{I_L(1)}, \cdots, f_{I_L(N_c-1)}]$.
4. The subcarrier frequency vector F_L is used to assign subcarriers to the users. Recall, for user $i = 0, 1 \cdots, M-1$, the total number of subcarriers assigned to the ith user is N_u^i. We assign subcarriers $\{f_{I_L(0)}, f_{I_L(1)}, \cdots, f_{I_L(N_u^0-1)}\}$ to user 0; Assign $\{f_{I_L(N_u^0)}, f_{I_L(N_u^0+1)}, \cdots, f_{I_L(N_u^0+N_u^1-1)}\}$ to user 1 and so on.

Remark 4.1 We would like to emphasize that *the secure subcarrier assignment scheme can be applied to any multiband, multiaccess communication systems and is not limited to OFDMA.*

Proposition 4.1 *The secure subcarrier assignment scheme ensures non-overlapping transmission among all the users in the system.*

Proof In fact, after L steps, we obtain the subcarrier frequency vector as $F_L = [f_{I_L(0)}, f_{I_L(1)}, \cdots, f_{I_L(N_c-1)}]$. We can rewrite the subcarrier frequency vector F_L as $F_L = [F_L(0), F_L(1), \cdots, F_L(N_c - 1)]$ by defining $F_L(j) = f_{I_L(j)}$ for $j = 0, 1, \cdots, N_c - 1$, where N_c is the total number of subcarriers. Assume that we have M users in the system, and for $i = 0, 1 \cdots, M - 1$, the total number of subcarriers assigned to the ith user is N_u^i. The subcarrier assignment process described in step 4 of the secure subcarrier algorithm above is equivalent to assigning subcarriers $\{F_L(0), F_L(1), \cdots, F_L(N_u^0 - 1)\}$ to user 0 and subcarriers $\{F_L(N_u^0), F_L(N_u^0 + 1), \cdots, F_L(N_u^0 + N_u^1 - 1)\}$ to user 1 and so on.

Because each frequency index appears in F_L once and only once, the algorithm ensures that (i) all the users are transmitting on non-overlapping sets of subcarriers and (ii) no subcarrier is left idle. That is, all the subcarriers are active. ■

The secure permutation index generation is performed at the base station. The base station sends encrypted channel assignment information to each user periodically through the control channels.

The secure subcarrier assignment scheme addresses the problem of securely allocating subcarriers in the presence of hostile jamming. This algorithm can be combined with existing resource allocation techniques. First, the number of subcarriers assigned to each user can be determined through power and bandwidth optimization; see [10, 17], for example. Then, we use the secure subcarrier assignment algorithm to select the group of subcarriers for each user at each hopping period. In the following, we illustrate the secure subcarrier assignment algorithm though a simple example.

Example Assume the total number of available subcarriers is $N_c = 8$, to be equally divided among $M = 2$ users, the permutation index vector $P = [4, 7, 4, 0]$, and the initial index vector $I_0 = [0, 1, 2, 3, 4, 5, 6, 7]$, as shown in Fig. 4.1. Note that, the initial index vector I_0 can contain any random permutation of the sequence $\{0, 1, \cdots, N_c - 1\}$, and $L \in [\frac{N_c}{2}, N_c]$. In this example, we choose $L = \frac{N_c}{2}$.

At Step 1, $k = 1$, and $P_k = 4$, thus we switch $I_0(P_k)$ and $I_0(k - 1)$ of the index vector I_0. After the switching, we obtain a new index vector $I_1 = [4, 1, 2, 3, 0, 5, 6, 7]$.

At Step 2, $k = 2$, and $P_k = 7$, thus we switch $I_1(P_k)$ and $I_1(k - 1)$ of the index vector I_1. We obtain the new index vector $I_2 = [4, 7, 2, 3, 0, 5, 6, 1]$. Below are the remaining index vectors for $k = 3, 4$:

$$I_3 = [4, 7, 0, 3, 2, 5, 6, 1], \ I_4 = [3, 7, 0, 4, 2, 5, 6, 1].$$

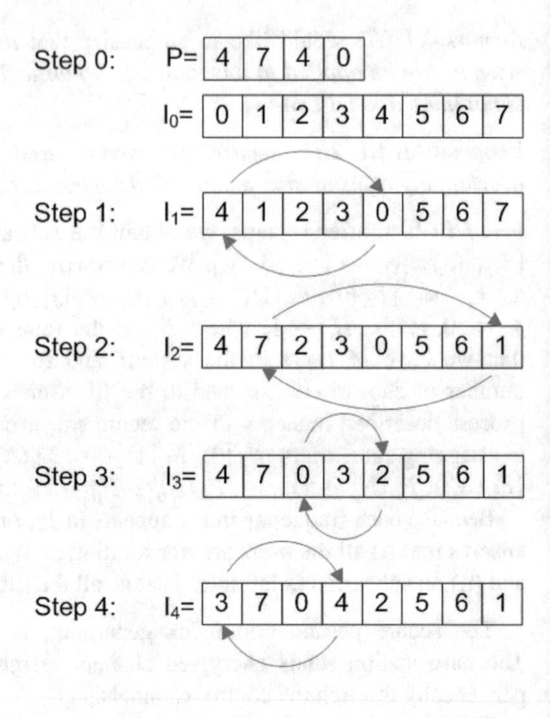

Fig. 4.1 Example of the secure permutation algorithm for $N_c = 8$ subcarriers and $M = 2$ users

The subcarrier frequency vector is $F_4 = [f_{I_4(0)}, f_{I_4(1)}, \cdots, f_{I_4(N_c-1)}]$. Frequencies $\{f_3, f_7, f_0, f_4\}$ are assigned to user 0, and frequencies $\{f_2, f_5, f_6, f_1\}$ are assigned to user 1.

In the following section, we will introduce the CFFH system.

4.3 The Collision-Free Frequency Hopping (CFFH) Scheme

The CFFH system is essentially an OFDMA system equipped with secure FH-based dynamic spectrum access control, where the hopping pattern is determined by the secure subcarrier assignment algorithm described in the previous section.

4.3.1 Signal Transmission

Consider a system with M users, utilizing an OFDM system with N_c subcarriers, $\{f_0, \cdots, f_{N_c-1}\}$. At each hopping period, each user is assigned a specific subset of the total available subcarriers. One hopping period may last one or more OFDM symbol periods. Assuming that at the nth symbol, user i has been assigned a set of sub-carriers $C_{n,i} = \{f_{n,i_0}, \cdots, f_{n,i_{N_u^i-1}}\}$, that is, user i will transmit and only transmit on these subcarriers. Here N_u^i is the total number of subcarrier assigned to user i. Note that for any n,

$$C_{n,i} \bigcap C_{n,j} = \emptyset, \quad \text{if} \quad i \neq j. \tag{4.4}$$

That is, users transmit on non-overlapping subcarriers. In other words, there is no collision between the users. Ideally, for full capacity of the OFDM system,

$$\bigcup_{i=0}^{M-1} C_{n,i} = \{f_0, \cdots, f_{N_c-1}\}. \tag{4.5}$$

For the ith user, if $N_u^i > 1$, then the ith user's information symbols are first fed into a serial-to-parallel converter. Assuming that at the nth symbol period, user i transmits the information symbols $\{u_{n,0}^{(i)}, \cdots, u_{n,N_u^i-1}^{(i)}\}$ (which are generally QAM symbols) through the subcarrier set $C_{n,i} = \{f_{n,i_0}, \cdots, f_{n,i_{N_u^i-1}}\}$. User i's transmitted signal at the nth OFDM symbol can then be written as:

$$s_n^{(i)}(t) = \sum_{l=0}^{N_u^i-1} u_{n,l}^{(i)} e^{j2\pi f_{n,i_l} t}. \tag{4.6}$$

Note that each user does not transmit on subcarriers which are not assigned to him/her, by setting the symbols to zeros over these subcarriers. This process ensures collision-free transmission among the users.

4.3.2 Signal Detection

At the receiver, the received signal is a superposition of the signals transmitted from all users

$$r(t) = \sum_{i=0}^{M-1} r_n^{(i)}(t) + n(t), \tag{4.7}$$

where

$$r_n^{(i)}(t) = s_n^{(i)}(t) * h_i(t), \tag{4.8}$$

and $n(t)$ is the additive noise. In (4.8), $h_i(t)$ is the channel impulse response corresponding to user i. Note that in OFDM systems, guard intervals are inserted between symbols to eliminate intersymbol interference (ISI), so it is reasonable to study the signals in a symbol-by-symbol manner. Equations (4.6), (4.7), and (4.8) represent an uplink system. The downlink system can be formulated in a similar manner.

As is well known, the OFDM transmitter and receiver are implemented through IFFT and FFT, respectively. Denote the $N_c \times 1$ symbol vector corresponding to user i's nth OFDM symbol as $\mathbf{u}_n^{(i)}$; we have

$$\mathbf{u}_n^{(i)}(l) = \begin{cases} 0, & l \notin \{i_0, \cdots, i_{N_u^i-1}\} \\ u_{n,l}^{(i)}, & l \in \{i_0, \cdots, i_{N_u^i-1}\}. \end{cases} \tag{4.9}$$

Let T_s denote the OFDM symbol period. The discrete form of the transmitted signal $s_n^{(i)}(t)$ (sampled at $\frac{lT_s}{N_c}$) is

$$\mathbf{s}_n^{(i)} = \mathbf{F}\mathbf{u}_n^{(i)}, \tag{4.10}$$

where $\mathbf{F}$ is the IFFT matrix defined as

$$\mathbf{F} = \frac{1}{\sqrt{N_c}} \begin{pmatrix} W_{N_c}^{00} & \cdots & W_{N_c}^{0(N_c-1)} \\ \vdots & \ddots & \vdots \\ W_{N_c}^{(N_c-1)0} & \cdots & W_{N_c}^{(N_c-1)(N_c-1)} \end{pmatrix},$$

with $W_{N_c}^{nk} = e^{j2\pi nk/N_c}$. As we only consider one OFDM symbol at a time, for notation simplification, here we omit the insertion of the guard interval (i.e., the cyclic prefix which is used to ensure that there is no ISI between two successive OFDM symbols).

Let $\mathbf{h}_i = [h_i(0), \cdots, h_i(N_c - 1)]$ be the discrete channel impulse response vector, and let

$$\mathbf{H}_i = \mathbf{F}\mathbf{h}_i \tag{4.11}$$

be the Fourier transform of $\mathbf{h}_i$. Then the received signal corresponding to user i is

$$\mathbf{r}_n^{(i)}(l) = \mathbf{u}_n^{(i)}(l)\mathbf{H}_i(l). \tag{4.12}$$

The overall received signal is then given by

$$\mathbf{r}_n(l) = \sum_{i=0}^{M-1} \mathbf{r}_n^{(i)}(l) + \mathbf{N}_n(l) \tag{4.13}$$

$$= \sum_{i=0}^{M-1} \mathbf{u}_n^{(i)}(l)\mathbf{H}_i(l) + \mathbf{N}_n(l). \tag{4.14}$$

where $\mathbf{N}_n(l)$ is the Fourier transform of the noise corresponding to the nth OFDM symbol.

Note that due to the collision-free subcarrier assignment, for each l, there is at most one nonzero item in the sum $\sum_{i=0}^{M-1} \mathbf{u}_n^{(i)}(l)\mathbf{H}_i(l)$. As a result, standard channel estimation algorithms and signal detection algorithms for OFDM systems can be implemented. In fact, each user can send pilot symbols on its subcarrier set to perform channel estimation. It should be pointed out that instead of estimating the whole frequency domain channel vector $\mathbf{H}_i$, for signal recovery, user i only need to estimate the entries corresponding to its subcarrier set, that is the values of $\mathbf{H}_i(l)$ for $l \in \{i_0, \cdots, i_{N_u^i-1}\}$. After channel estimation, user i's information symbols can be estimated from

$$\mathbf{u}_n^{(i)}(l) = \frac{\mathbf{r}_n^{(i)}(l)}{\mathbf{H}_i(l)}, \quad l \in \{i_0, \cdots, i_{N_u^i-1}\}. \tag{4.15}$$

It is also interesting to note that we can obtain adequate channel information from all the users simultaneously, which can be exploited for dynamic resource reallocation to achieve better BER performance and real-time jamming prevention.

4.4 Space-Time Coded Collision-Free Frequency Hopping

In this section, we consider to enhance the anti-jamming features of the CFFH scheme using space-time coding. Here we present the transmitter and receiver design of the STC-CFFH system from the downlink perspective. The uplink can be designed in a similar manner.

4.4.1 Transmitter Design

We assume that during each hopping period, the number of subcarriers assigned to each user in the CFFH system is fixed. Recall that one hopping period may contain one or more OFDM symbol periods. In the following we illustrate the transmitter design over one OFDM symbol.

Assume the transmitter at the base station has n_T antennas and there are M users in the system. Over each OFDM symbol period, the ith user is assigned N_u^i subcarriers, which do not need to be contiguous. The transmitter structure at the base station is illustrated in Fig. 4.2.

Initially, the input bit stream corresponding to each user is mapped to symbols based on a selected constellation. The constellation could be different for different users based on the channel condition and user data rate [18, 19]. Assume the base station uses a $n_T \times n_T$ space-time block code (STBC). Note that non-square STBC codes [14, 15] exist, but for notation simplicity, here we adopt the $n_T \times n_T$ square

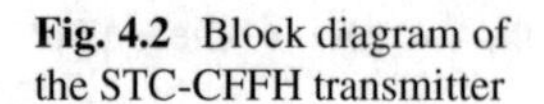

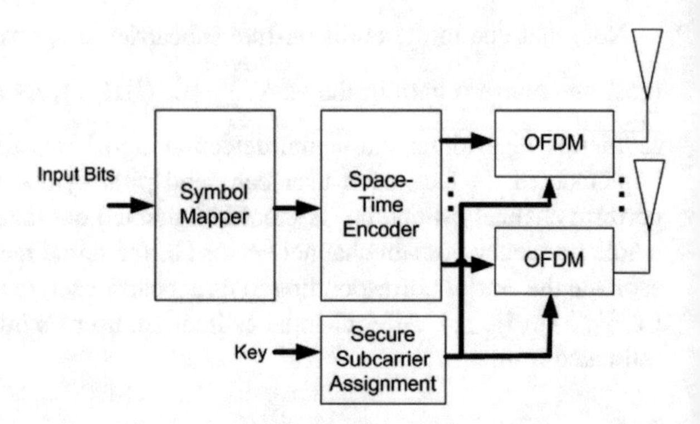

Fig. 4.2 Block diagram of the STC-CFFH transmitter

code. For each user, divide the N_u^i subcarriers into $G_i = \frac{N_u^i}{n_T}$ groups, where each group contains n_T subcarriers, which is of the same length as that of the STBC. For simplicity, we assume G_i is an integer, that is, each user transmits G_i space-time blocks in one OFDM symbol period (otherwise, if G_i is not an integer, the symbols can be broken down and transmitted over two successive OFDM symbol periods).

For each $n \in \{1, 2, \cdots, G_i\}$, the base station takes a block of n_T complex symbols and maps them to a $n_T \times n_T$ STBC code matrix $X_i(n)$. In other words, for $n = 1, 2, \cdots, G_i$, $m = 1, 2, \cdots, n_T$, the mth row of the code matrix $X_i(n)$ is merged with the corresponding symbols from other users and transmitted through the mth transmit antenna, and all symbols within each column ($m = 1, 2, \cdots, n_T$) of the code matrix $X_i(n)$ is transmitted over the same subcarrier. The code matrix $X_i(n)$ is given by

$$X_i(n) = \overset{\text{Subcarrier} \rightarrow}{\begin{bmatrix} x_{i,1}^1(n) & \cdots & x_{i,n_T}^1(n) \\ \vdots & \ddots & \vdots \\ x_{i,1}^{n_T}(n) & \cdots & x_{i,n_T}^{n_T}(n) \end{bmatrix}} \downarrow \text{Antenna}, \tag{4.16}$$

where $x_{i,t}^m(n)$ is the tth symbol of the nth block for user i in transmit antenna m.

Note that, since each user is assigned multiple frequency bands, we are transmitting symbols over multiple subcarriers instead of multiple time slots. Thus the time diversity of the space-time coder is converted to frequency diversity, and this structure is referred to as space-frequency coding [20].

STC-CFFH Transmitter Design Example We provide an example to illustrate the transmitter structure of STC-CFFH, in which the subcarrier assignment is based on the example in Sect. 4.2. Assume an Alamouti space-time coded system with n_T=2, and we have M=2 users. A total of N_c=8 subcarriers are available, and each user is assigned $N_u^0 = N_u^1 = 4$ subcarriers. For this example, each user transmit $G_i = \frac{N_u^i}{n_T} = 2$ code matrices in one OFDM symbol period. Consider the nth block for the ith user, where $n = 1, 2$ in this case. The space-time encoder takes n_T=2 complex symbols $x_{i,1}(n)$, $x_{i,2}(n)$ in each encoding operation and maps them to the code matrix $X_i(n)$. In this example, the first and second row of $X_i(n)$ will be sent from the first and second transmit antennas, respectively.

In this example, we can drop the superscript m in $x_{i,t}^m(n)$ by representing $X_i(n)$ with the Alamouti space-time code block structure [13]. Then the code matrices $X_i(n)$ are given by

$$X_i(n) = \overset{\text{Subcarrier} \rightarrow}{\begin{bmatrix} x_{i,1}(n) & -x_{i,2}^*(n) \\ x_{i,2}(n) & x_{i,1}^*(n) \end{bmatrix}} \downarrow \text{Antenna}, \tag{4.17}$$

where * is the complex conjugate operator. Specifically, user 0's two code matrices are represented as

$$X_0(1) = \begin{bmatrix} x_{0,1}(1) & -x_{0,2}^*(1) \\ x_{0,2}(1) & x_{0,1}^*(1) \end{bmatrix}, \ X_0(2) = \begin{bmatrix} x_{0,1}(2) & -x_{0,2}^*(2) \\ x_{0,2}(2) & x_{0,1}^*(2) \end{bmatrix}, \tag{4.18}$$

and user 1's two code matrices are represented as

$$X_1(1) = \begin{bmatrix} x_{1,1}(1) & -x_{1,2}^*(1) \\ x_{1,2}(1) & x_{1,1}^*(1) \end{bmatrix}, \ X_1(2) = \begin{bmatrix} x_{1,1}(2) & -x_{1,2}^*(2) \\ x_{1,2}(2) & x_{1,1}^*(2) \end{bmatrix}. \tag{4.19}$$

Recall the secure subcarrier assignment from the example in Sect. 4.2. User 0 is assigned to subcarriers $\{f_0, f_3, f_4, f_7\}$. User 1 is assigned to subcarriers $\{f_1, f_2, f_5, f_6\}$. A depiction of the subcarrier allocation for this example is provided in Table 4.1.

For user 0, $[x_{0,1}(1), -x_{0,2}^*(1), x_{0,1}(2), -x_{0,2}^*(2)]$ is transmitted through antenna 1 over subcarriers $\{f_0, f_3, f_4, f_7\}$, respectively; $[x_{0,2}(1), x_{0,1}^*(1), x_{0,2}(2), x_{0,1}^*(2)]$ is transmitted through antenna 2 over the same group of subcarriers. User 1's subcarrier allocation can be achieved in the same manner as User 0.

Table 4.1 STC-CFFH transmitter example

freq. / *Tx.*	f_0	f_1	f_2	f_3	f_4	f_5	f_6	f_7
1	$x_{0,1}(1)$	$x_{1,1}(1)$	$-x_{1,2}^*(1)$	$-x_{0,2}^*(1)$	$x_{0,1}(2)$	$x_{1,1}(2)$	$-x_{1,2}^*(2)$	$-x_{0,2}^*(2)$
2	$x_{0,2}(1)$	$x_{1,2}(1)$	$x_{1,1}^*(1)$	$x_{0,1}^*(1)$	$x_{0,2}(2)$	$x_{1,2}(2)$	$x_{1,1}^*(2)$	$x_{0,1}^*(2)$

4.4.2 Receiver Design

Assume user i has n_R antennas. Recall that the secure permutation index generation is performed at the base station and the base station sends encrypted channel assignment information to each user periodically through the control channels. After cyclic prefix removal and FFT, the receiver will only extract the symbols on the subcarriers assigned to itself and discard the symbols on the rest of subcarriers. The extracted symbols are reorganized into a $n_R \times n_T$ matrix $R_i(n)$, which corresponds to the transmitted code matrix $X_i(n)$. Thus the space-time decoding can be performed for each symbol matrix $R_i(n)$ individually, and the estimated symbols are mapped back into bits by the symbol demapper.

Here we consider the space-time decoding algorithm for a single symbol matrix $R_i(n)$ given as

$$R_i(n) = \overset{\text{Subcarrier} \rightarrow}{\begin{bmatrix} r_{i,1}^{1}(n) & \cdots & r_{i,n_T}^{1}(n) \\ \vdots & \ddots & \vdots \\ r_{i,1}^{n_R}(n) & \cdots & r_{i,n_T}^{n_R}(n) \end{bmatrix}} \downarrow \text{Antenna}, \tag{4.20}$$

where $r_{i,t}^{j}(n)$ is the tth symbol of group n for user i from jth receive antenna. Each symbol in the matrix $R_i(n)$ can be obtained as

$$r_{i,t}^{j}(n) = \sum_{m=1}^{n_T} H_{i,t}^{j,m}(n) x_{i,t}^{m}(n) + n_{i,t}^{j}(n), \tag{4.21}$$

where $H_{i,t}^{j,m}(n)$ is the channel frequency response for the path from the mth transmit antenna to the jth receive antenna corresponding to the tth symbol of group n for user i. It is assumed that the channels between the different antennas are uncorrelated. Here, $n_{i,t}^{j}(n)$ is the OFDM-demodulated version of the additive white Gaussian noise (AWGN) at the jth receive antenna for the tth symbol of the nth group for the ith user. The noise is assumed to be zero mean with variance σ_N^2.

The space-time maximum likelihood (ML) decoder is obtained as

$$\hat{X}_i(n) = \arg\min_{X_i(n)} \sum_{j=1}^{n_R} \sum_{t=1}^{n_T} \left| r_{i,t}^{j}(n) - \sum_{m=1}^{n_T} H_{i,t}^{j,m}(n) x_{i,t}^{m}(n) \right|^2, \tag{4.22}$$

where $\hat{X}_i(n)$ denotes the recovered symbols of group n for user i. Note that the minimization is performed over all possible space-time code words.

STC-CFFH Receiver Design Example We continue with the transmitter example in the previous subsection. Assuming each user is equipped with $n_R = 2$ receive antennas, the received symbols are illustrated in Table 4.2. Arranging the extracted

Table 4.2 STC-CFFH receiver example

Rx. \ freq.	f_0	f_1	f_2	f_3	f_4	f_5	f_6	f_7
1	$r_{0,1}^1(1)$	$r_{1,1}^1(1)$	$r_{1,2}^1(1)$	$r_{0,2}^1(1)$	$r_{0,1}^1(2)$	$r_{1,1}^1(2)$	$r_{1,2}^1(2)$	$r_{0,2}^1(2)$
2	$r_{0,1}^2(1)$	$r_{1,1}^2(1)$	$r_{1,2}^2(1)$	$r_{0,2}^2(1)$	$r_{0,1}^2(2)$	$r_{1,1}^2(2)$	$r_{1,2}^2(2)$	$r_{0,2}^2(2)$

symbols according to the users and the groups, the extracted symbol matrix $R_i(n)$ is given as

$$R_i(n) = \overset{\text{Subcarrier} \rightarrow}{\begin{bmatrix} r_{i,1}^1(n) \; r_{i,2}^1(n) \\ r_{i,1}^2(n) \; r_{i,2}^2(n) \end{bmatrix}} \downarrow \text{Antenna}. \tag{4.23}$$

Specifically, user 0's two extracted symbol matrices can be represented as

$$R_0(1) = \begin{bmatrix} r_{0,1}^1(1) \; r_{0,2}^1(1) \\ r_{0,1}^2(1) \; r_{0,2}^2(1) \end{bmatrix}, \; R_0(2) = \begin{bmatrix} r_{0,1}^1(2) \; r_{0,2}^1(2) \\ r_{0,1}^2(2) \; r_{0,2}^2(2) \end{bmatrix}, \tag{4.24}$$

and user 1's two extracted symbol matrices can be represented as

$$R_1(1) = \begin{bmatrix} r_{1,1}^1(1) \; r_{1,2}^1(1) \\ r_{1,1}^2(1) \; r_{1,2}^2(1) \end{bmatrix}, \; R_1(2) = \begin{bmatrix} r_{1,1}^1(2) \; r_{1,2}^1(2) \\ r_{1,1}^2(2) \; r_{1,2}^2(2) \end{bmatrix}. \tag{4.25}$$

Then, the ML space-time decoding is performed for each $R_i(n)$.

Remark 4.2 In the discussion above, we focused on STC-CFFH system for the downlink case, where the information is transmitted from base station to the multiple users. In the uplink case, the secure permutation index is encrypted and transmitted from base station to each user, prior to the user transmission. Then during the transmission, each user only transmits on the subcarriers assigned to him/her. The receiver at the base station separates each user's transmitted data. In order for the user to use space-time coding, each user needs to have at least two antennas.

4.5 Performance Analysis of STC-CFFH

In this section, we investigate the spectral efficiency and the performance of the STC-CFFH schemes under jamming interference over frequency-selective fading environments. First, the system performance in jamming-free case is analyzed. Second, the system performance under hostile jamming is investigated. Finally, the spectral efficiency comparison of STC-CFFH and the conventional FH-OFDMA system is performed.

4.5.1 System Performance in Jamming-Free Case

First, we analyze the pairwise error probability of the STC-CFFH system under Rayleigh fading. Assume ideal channel state information (CSI) and perfect synchronization between transmitter and receiver. Recall that the ML space-time decoding rule for the extracted symbol matrix $R_i(n)$ is given by (4.22).

Denote the pairwise error probability of transmitting $X_i(n)$ and deciding in favor of another code word $\hat{X}_i(n)$, given the realizations of the fading channel $H_{i,t}^{j,m}(n)$, as $P(X_i(n), \hat{X}_i(n)|H_{i,t}^{j,m}(n))$. This pairwise error probability is bounded by [21] (see page 255)

$$P(X_i(n), \hat{X}_i(n)|H_{i,t}^{j,m}(n)) \leq \exp\left(-d^2(X_i(n), \hat{X}_i(n))\frac{E_s}{4N_0}\right), \tag{4.26}$$

where E_s is the average symbol energy, N_0 is the noise power spectral density, and $d^2(X_i(n), \hat{X}_i(n))$ is a modified Euclidean distance between the two space-time code words $X_i(n)$ and $\hat{X}_i(n)$ and is given by

$$d^2(X_i(n), \hat{X}_i(n)) = \sum_{t=1}^{n_T}\sum_{j=1}^{n_R}\left|\sum_{m=1}^{n_T} H_{i,t}^{j,m}(n)(\hat{x}_{i,t}^m(n) - x_{i,t}^m(n))\right|^2, \tag{4.27}$$

where $\hat{x}_{i,t}^m(n)$ is the estimated version of $x_{i,t}^m(n)$.

Let us define a code word difference matrix $C(X_i(n), \hat{X}_i(n)) = X_i(n) - \hat{X}_i(n)$ and define a code word distance matrix $B(X_i(n), \hat{X}_i(n))$ with rank r_B as

$$B(X_i(n), \hat{X}_i(n)) = C(X_i(n), \hat{X}_i(n)) \cdot C(X_i(n), \hat{X}_i(n))^H, \tag{4.28}$$

where H denotes the Hermitian operator. Since the matrix $B(X_i(n), \hat{X}_i(n))$ is a nonnegative definite Hermitian matrix, the eigenvalues of $B(X_i(n), \hat{X}_i(n))$ are nonnegative real numbers, denoted as $\lambda_1, \lambda_2, \cdots, \lambda_{r_B}$.

After averaging with respect to the Rayleigh fading coefficients, the upper bound of pairwise error probability can be obtained as [22]

$$P(X_i(n), \hat{X}_i(n)|H_{i,t}^{j,m}(n)) \leq \left(\prod_{j=1}^{r_B}\lambda_j\right)^{-n_R}\left(\frac{E_s}{4N_0}\right)^{-r_B n_R}. \tag{4.29}$$

In the case of low signal-to-noise ratio (SNR), the upper bound in (4.29) can be expressed as [21]

$$P(X_i(n), \hat{X}_i(n)|H_{i,t}^{j,m}(n)) \leq \left(1 + \frac{E_s}{4N_0}\sum_{j=1}^{r_B}\lambda_j\right)^{-n_R}. \tag{4.30}$$

4.5.2 *System Performance Under Hostile Jamming*

In this subsection, we will first introduce the jamming models and then analyze the system performance under both full-band jamming and partial-band jamming.

4.5.2.1 Jamming Models

Jamming interference in the OFDM framework can severely degrade the system performance [23]. Each extracted symbol in the matrix $R_i(n)$ that experiences jamming interference is given as

$$r^j_{i,t}(n) = \sum_{m=1}^{n_T} H^{j,m}_{i,t}(n) x^m_{i,t}(n) + n^j_{i,t}(n) + J^j_{i,t}(n), \tag{4.31}$$

where $J^j_{i,t}(n)$ is the jamming interference at the jth receive antenna for the tth symbol of the nth group for the ith user. Assume all jamming interference $J^j_{i,t}(n)$ has the same power spectral density N_J, then the signal-to-jamming plus noise ratio (SJNR) at the receiver is represented by $\text{SJNR} = \frac{E_s}{N_0+N_J}$. When the noise is dominated by jamming, the SJNR can be represented as the signal-to-jamming ratio (SJR) where $\text{SJR} = \frac{E_s}{N_J}$.

Partial-band jamming [24–26] is generally characterized by the additive Gaussian noise interference with flat power spectral density $\frac{N_J}{\rho}$ over a fraction ρ of the total bandwidth and negligible interference over the remaining fraction $(1 - \rho)$ of the band. ρ is also referred to as the jammer occupancy and is given as

$$\rho = \frac{W_J}{W_S} \leq 1, \tag{4.32}$$

where W_J is the jamming bandwidth and W_S is the total signal bandwidth. For CFFH, partial-band jamming means that the jamming power is concentrated on a certain group of subcarriers. Let n_J denotes the number of jammed subcarriers, and then the jamming ratio ρ is given by $\rho = \frac{n_J}{n_T}$. For a particular code matrix $X_i(n)$, this means that on average, ρn_T subcarriers are jammed out of n_T subcarriers used by $X_i(n)$.

When $\rho = 1$, the jamming power is uniformly distributed over the entire bandwidth. In this case, the partial-band jamming becomes *full-band jamming* [27, 28]. For a CFFH system, full-band jamming means that the jamming power is uniformly distributed over all N_c.

4.5.2.2 System Performance Under Rayleigh Fading and Full-Band Jamming

In the presence of Rayleigh fading and full-band jamming, the pairwise error probability can be expressed in terms of the jamming power spectral density N_J and average signal power E_s. In the case of high SNR, the upper bound in (4.29) can be expressed as

$$P(X_i(n), \hat{X}_i(n)|H_{i,t}^{j,m}(n)) \leq \left(\prod_{j=1}^{r_B} \lambda_j\right)^{-n_R} \left(\frac{E_s}{4N_J}\right)^{-r_B n_R}. \tag{4.33}$$

From (4.30), the upper bound in the presence of Rayleigh fading and full-band jamming can be expressed as

$$P(X_i(n), \hat{X}_i(n)|H_{i,t}^{j,m}(n)) \leq \left(1 + \frac{E_s}{4(N_0+N_J)} \sum_{j=1}^{r_B} \lambda_j\right)^{-n_R}. \tag{4.34}$$

As will be confirmed in Sect. 4.6, for the STC-CFFH system, the space-frequency diversity gain is insignificant at low SJNR; however, the diversity gain becomes noticeable at high SJNR.

4.5.2.3 System Performance Under Rayleigh Fading and Partial-Band Jamming

Recall that each column of the received symbol matrix $R_i(n)$ is obtained from the same subcarrier in all receive antennas. When we have partial-band jamming, most likely not all columns of $R_i(n)$ are jammed, since each column is transmitted through a different subcarrier. Thus the receiver may be able to recover the transmitted signal relying on the jamming-free columns.

Orthogonal space-time codes (OSTC) are capable of perfectly decoding the transmitted symbols under partial-band jamming and noise-free environments when at least one frequency band is not jammed. We consider a $n_T = 4$ space-time orthogonal block code design as an example. Following the same notation convention in the STC-CFFH transmitter example in Sect. 4.4, the code matrix with transmit symbols $x_{i,t}(n)$ for $t = 1, 2, 3, 4$, is represented as

$$X_i(n) = \begin{bmatrix} x_{i,1}(n) & x_{i,2}(n) & x_{i,3}(n) & x_{i,4}(n) \\ -x_{i,2}(n) & x_{i,1}(n) & -x_{i,4}(n) & x_{i,3}(n) \\ -x_{i,3}(n) & x_{i,4}(n) & x_{i,1}(n) & -x_{i,2}(n) \\ -x_{i,4}(n) & -x_{i,3}(n) & x_{i,2}(n) & x_{i,1}(n) \end{bmatrix}. \tag{4.35}$$

Due to the orthogonality of the code design, each frequency band contains full information about the transmitted symbols. As a result, the transmitted symbols are recovered perfectly when there is at least one unjammed frequency band.

In this case, the average probability of error P_e can be expressed as

$$P_e = \sum_{i=0}^{4} P_{e,i} \; Pr\{i \; out \; of \; 4 \; bands \; are \; jammed\}, \tag{4.36}$$

where $P_{e,i}$ is the probability of error when i out of 4 bands are jammed.

4.5.3 Spectral Efficiency

One major challenge in the current FH-OFDMA system is collision. In FH-OFDMA, multiple users hop their subcarrier frequencies independently. If two users transmit simultaneously in the same frequency band, a collision, or hit occurs. In this case, the probability of bit error is generally assumed to be 0.5 [29].

If there are N_c available channels and M active users (i.e., $M - 1$ possible interfering users), assuming that all N_c channels are equally probable and all users are independent, even if each user only transmits over a single carrier, then the probability that a collision occurs is given by

$$P_h = 1 - \left(1 - \frac{1}{N_c}\right)^{M-1} \tag{4.37}$$

$$\approx \frac{M-1}{N_c} \quad \text{when } N_c \text{ is large.} \tag{4.38}$$

The high collision probability severely limits the number of users that can be simultaneously supported by an FH-OFDMA system.

Take $N_c = 64$ as an example. For a required BER of 0.04, only six users can be supported. That is, only 6 out of 64 subcarriers can be used simultaneously; the carrier efficiency is $\frac{6}{64} = 9.38\%$. On the other hand, due to the collision-free design, CFFH has the same spectral efficiency and BER performance as that of OFDM. For CFFH, the carrier efficiency is 100% with a much better BER performance. In this particular case, CFFH is approximately 10.67 times more efficient than the conventional FH-OFDMA system. This fact is further illustrated in simulation example 1 of Sect. 4.6.

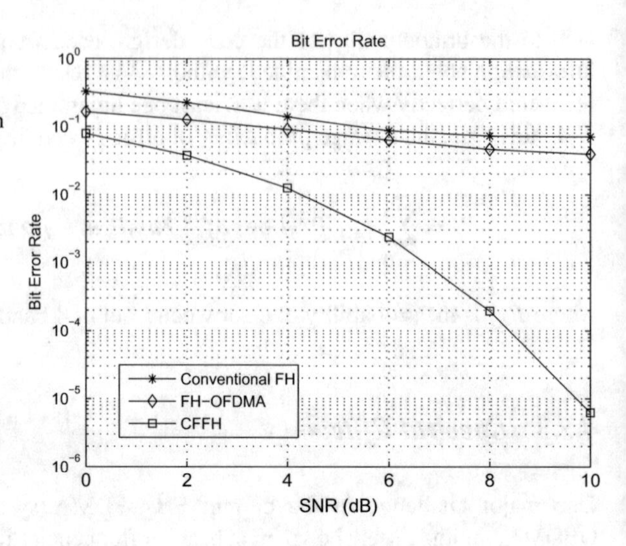

Fig. 4.3 BER performance over AWGN channel of the CFFH, FH-OFDMA, and the conventional FH systems with M = 8 users and $N_c = 128$ available subcarriers

4.6 Simulation Examples

In this section, we provide simulation examples to demonstrate the performance of the CFFH and STC-CFFH schemes. First, the bit error performances of the CFFH scheme, the conventional FH, and FH-OFDMA systems are compared under AWGN channels. Second, the bit error performance of the CFFH scheme, the STC-CFFH scheme, and the STC-OFDM system is performed over a frequency-selective fading channel with partial-band jamming.

Simulation example 1 We consider the conventional FH, the FH-OFDMA, and the CFFH systems, each with M = 8 users and $N_c = 128$ available subcarriers. The conventional FH system uses four-frequency shift keying (4-FSK) modulation, where each user transmits over a single carrier. Both the CFFH and FH-OFDMA systems transmit 16-QAM symbols, and each user is assigned 16 subcarriers. The average bit error rate (BER) versus the signal-to-noise ratio (SNR) performance over AWGN channels of the systems is illustrated in Fig. 4.3. As can be seen, the CFFH scheme delivers excellent results since the multiuser access interference (MAI) is avoided. The conventional FH and FH-OFDMA schemes, on the other hand, are severely limited by collision effect among users.

Simulation example 2 The BER performances of the STC-OFDM scheme and the STC-CFFH and CFFH schemes are evaluated by simulations. The simulations are carried out over a frequency-selective Rayleigh fading channel with partial-band jamming. An Alamouti space-time coding system with two transmit and receive antennas is applied to the STC-CFFH system. We assume perfect timing and frequency synchronization, as well as uncorrelated channels for each antenna. The total number of available subcarriers is $N_c = 256$, and the number of users is M = 16; therefore each user is assigned 16 subcarriers.

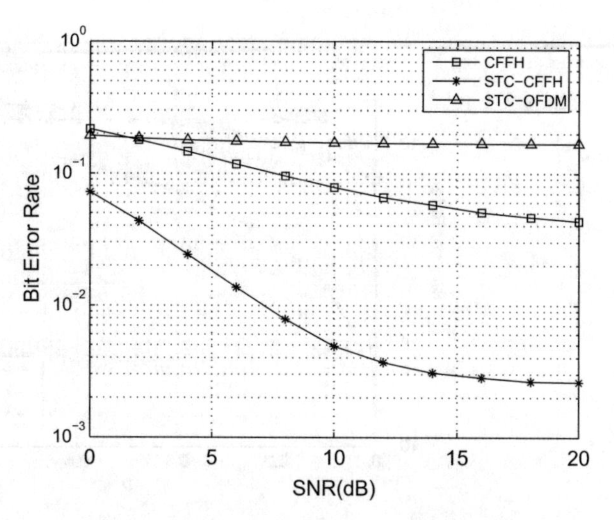

Fig. 4.4 Comparison of the BER over frequency-selective fading channel with partial-band jamming. Number of subcarriers $N_c = 256$, number of users = 16 and SJR = 0 dB

We consider the performance of three systems that transmit 16-QAM symbols: (i) the CFFH system, (ii) an STC-OFDM system, and (iii) the STC-CFFH system. For system (ii), each user transmits on 16 fixed subcarriers. In systems (i) and (iii), each user transmits on 16 pseudorandom secure subcarriers. We assume the jammer intentionally interferes 16 subcarriers out of the whole band.

Figure 4.4 depicts the BER versus SNR over frequency-selective fading with SJR=0dB. Due to the secure subcarrier assignment, the CFFH system outperforms the STC-OFDM system. The pseudorandom secure subcarrier assignment randomizes each users' subcarrier occupancy (i.e., spectrum occupancy) at a given time, therefore allowing for multiple access over a wide range of frequencies. Furthermore, incorporating space-time coding into CFFH significantly increases the BER performance. We also notice that at high SNR levels, the performance-limiting factor for all systems is the partial-band jamming. In Fig. 4.5, the BER versus the jammer occupancy (ρ) is evaluated with SNR=10dB and SJR=0dB for the three systems. Recall the jammer occupancy is the fraction of subcarriers that experience interference. We can see that the STC-CFFH system outperforms the other systems for all $\rho < 1$. This example shows that STC-CFFH is very robust under jamming interference.

We also observe that due to the randomness in the frequency hopping pattern, as well as the fact that the system ensures collision-free transmission among the users, the performance of CFFH remains the same as the number of users varies in the system.

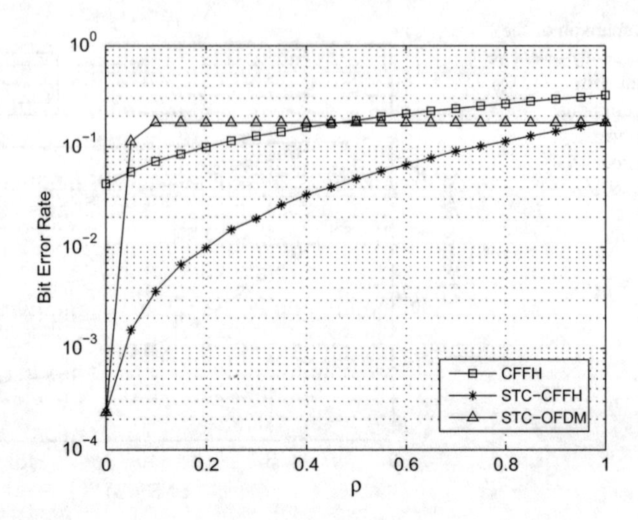

Fig. 4.5 BER versus jammer occupancy over frequency-selective fading channel with partial-band to full-band jamming. Number of subcarriers $N_c = 256$, number of users = 16, SJR = 0dB, and SNR = 10 dB

4.7 Conclusions

In this chapter, we introduced a secure collision-free frequency hopping scheme. Based on the OFDMA framework and the secure subcarrier assignment algorithm, the CFFH system can achieve high spectral efficiency through collision-free multiple access. While keeping the inherent anti-jamming, anti-interception security features of the FH system, CFFH can achieve the same spectral efficiency as that of OFDM and can relax the strict synchronization requirement suffered by the conventional FH systems. Furthermore, we enhanced the jamming resistance of the CFFH scheme by incorporating space-time coding to the system. The OFDMA-based dynamic spectrum access control scheme presented in this chapter can be applied directly for secure and efficient spectrum sharing among different users and services in wireless networks.

References

1. C. Shahriar, M. La Pan, M. Lichtman, T. C. Clancy, R. McGwier, R. Tandon, S. Sodagari, and J. H. Reed. Phy-layer resiliency in OFDM communications: A tutorial. *IEEE Communications Surveys Tutorials*, 17(1):292–314, Firstquarter 2015.
2. F. Dominique and J.H. Reed. Robust frequency hop synchronisation algorithm. *Electronics Letters*, 32:1450–1451, August 1996.
3. M. Simon, G. Huth, and A. Polydoros. Differentially coherent detection of QASK for frequency-hopping systems–Part I: Performance in the presence of a Gaussian noise environment. *IEEE Transactions on Communications*, 30:158–164, January 1982.

4. Y.M. Lam and P.H. Wittke. Frequency-hopped spread-spectrum transmission with band-efficient modulations and simplified noncoherent sequence estimation. *IEEE Transactions on Communications*, 38:2184–2196, December 1990.
5. Joonyoung Cho, Youhan Kim, and Kyungwhoon Cheun. A novel FHSS multiple-access network using M-ary orthogonal Walsh modulation. In *Proc. 52nd IEEE Veh. Technol. Conf.*, volume 3, pages 1134–1141, Sept. 2000.
6. S. Glisic, Z. Nikolic, N. Milosevic, and A. Pouttu. Advanced frequency hopping modulation for spread spectrum WLAN. *IEEE Journal on Selected Areas in Communications*, 18:16–29, January 2000.
7. Kwonhue Choi and Kyungwhoon Cheun. Maximum throughput of FHSS multiple-access networks using MFSK modulation. *IEEE Transactions on Communications*, 52:426–434, March 2004.
8. Kang-Chun Peng, Chien-Hsiang Huang, Chien-Jung Li, and Tzyy-Sheng Horng. High-performance frequency-hopping transmitters using two-point Delta-Sigma modulation. *IEEE Transactions on Microwave Theory and Techniques*, 52:2529–2535, November 2004.
9. K. Choi and K. Cheun. Optimum parameters for maximum throughput of FHMA system with multilevel FSK. *IEEE Transactions on Vehicular Technology*, 55:1485–1492, Sept. 2006.
10. C.Y. Wong, R.S. Cheng, K.B. Letaief, and R.D. Murch. Multiuser OFDM with adaptive subcarrier, bit and power allocatioin. *IEEE Journal of Selective Areas on Communications*, Oct 1999.
11. H. Sari. *Orthogonal frequency-division multiple access with frequency hopping and diversity*, pages 57–68. in *Multi-Carrier Spread Spectrum*, K. Fazel and G. P. Fettweis, Editors, Norwell, USA: Kluwer, 1997.
12. US National Institute of Standards and Technology. *Federal Information Processing Standards Publication 197-Announcing the* ADVANCE ENCRYPTION STANDARD (AES). http://csrc.nist.gov/publications/fips/fips197/fips-197.pdf, 2001.
13. S. Alamouti. A simple transmit diversity technique for wireless communications. *IEEE Journal on Selected Areas in Communications*, pages 1451–1458, October 1998.
14. V. Tarokh, H. Jafarkhani, and A.R. Calderbank. Space-time block code from orthogonal designs. *IEEE Trans. Information Theory*, July 1999.
15. V. Tarokh, H. Jafarkhani, and A. Calderbank. Space-time codes for high data rate wireless communication: Performance results. *IEEE Journal on Select Areas Communication*, pages 451–460, March 1999.
16. W.E. Burr. Selecting the advance encryption standard. *IEEE Security and Privacy*, pages 43–52, April 2003.
17. J. Jang and K.B. Lee. Transmit power adaptation for multiuser OFDM systems. *IEEE Journal of Selective Areas on Communications*, Feb. 2003.
18. M. Ergen, S. Coleri, and P. Varaiya. QoS aware adaptive resource allocation techniques for fair scheduling in OFDMA based broadband wireless access systems. *in Proceedings of IEEE Transactions on Broadcasting*, pages 362–370, Dec. 2003.
19. S.E. Elayoubi and B. Fourestie. Performance evaluation of admission control and adaptive modulation in OFDMA WiMax systems. *in Proceedings of IEEE/ACM Transactions on Networking*, pages 1200–1211, Oct. 2008.
20. K. Lee and D.B. Williams. A space-frequency transmitter diversity technique for OFDM systems. *in Proceedings of IEEE Global Communications Conference*, pages 1473–1477, Nov. 2000.
21. B. Vucetic and J. Yuan. *Space Time Coding*, chapter Space Time Coding Peformance Analysis and Code Design, page 2003. John Wiley and Sons, Ltd.
22. V. Tarokh, N. Seshadri, and A. Calderbank. Space-time codes for high data rate wireless communication: Performance criterion and code construction. *IEEE Trans. Information Theory*, pages 744–765, March 1998.
23. J. Park, D. Kim, C. Kang, and D. Hong. Effect of partial band jamming on OFDM-based WLAN in 802.11g. *IEEE International Conference on Acoustics, Speech, and Signal Processing*, pages 560–563, April 2003.

24. P.J. Crepeau. Performance of FH/BFSK with generalized fading in worst case partial-band Gaussian interference. *IEEE Journal on Selected Areas in Communications*, 8:884–886, June 1980.
25. M.B. Pursley and W.E. Stark. Performance of Reed-Solomon coded frequency-hop spread-spectrum communications in partial-band interference. *IEEE Transactions on Communications*, 33:767–774, August 1985.
26. W.E. Stark. Coding for frequency-hopped spread-spectrum communication with partial-band interference-Part II: Coded performance. *IEEE Transactions on Communications*, 33:1045–1057, October 1985.
27. R.L. Pickholtz, D.L. Schilling, and L.B.Milstein. Theory of spread spectrum communications - a tutorial. *IEEE Transactions on Communications*, 30:855–884, May 1982.
28. C.E. Cook and H.S. Marsh. An introduction to spread spectrum. *IEEE Communications Magazine*, 21:8–16, March 1983.
29. T.S. Rappaport. *Wireless Communications*. Prentice Hall, 2nd edition, 2002.

Chapter 5
Securely Precoded OFDM

5.1 Introduction

As is well known, due to its high spectral efficiency and robustness under fading channels, orthogonal frequency division multiplexing (OFDM) has been widely used in modern high-speed multimedia communication systems, such as LTE and WiMax [1]. However, unlike the spread spectrum techniques [2] that achieve jamming resistance through rich frequency diversity, OFDM mainly relies on channel coding for communication reliability under hostile jamming and has very limited built-in resilience against jamming attacks [3–9]. In [3], the bit error rate (BER) performance of the traditional OFDM was explored under full-band and partial-band Gaussian jamming, as well as multi-tone jamming. It was shown that OFDM is quite fragile under jamming, as the BER can go above 10^{-1} when the jamming power is the same as the signal power. In [6–8], jamming attacks aiming at the pilots in OFDM systems were studied. It was shown that when the system standard is public and no encryption is applied to the transmitted symbol sequence, pilot attacks can completely nullify the synchronization and channel estimation process of OFDM, hence resulting in complete communication failure.

In [5], the anti-jamming performance of frequency-hopped (FH) OFDM system was explored. Like the traditional FH system, this approach achieves jamming resistance through large frequency diversity and sacrifices the spectral efficiency of OFDM. In Chap. 4 of this book, a collision-free frequency hopping (CFFH) scheme [9] was presented, where the basic idea was to randomize the jamming interference through frequency domain interleaving based on secure, collision-free frequency hopping. *The most significant feature of CFFH based OFDM is that it is very effective under partial-band jamming and, at the same time, has the same spectral efficiency as the original OFDM.* However, CFFH-based OFDM is still fragile under *disguised jamming* [10–13].

T. Li et al., *Wireless Communications Under Hostile Jamming: Security and Efficiency*, https://doi.org/10.1007/978-981-13-0821-5_5

To combat disguised jamming in OFDM systems, a precoding scheme was proposed in [12], where extra redundancy is introduced to achieve jamming resistance. However, lack of plasticity in the precoding scheme results in inadequate reliability under cognitive disguised jamming. With OFDM being identified as a major modulation technique for the 5G systems, there is an ever-increasing need on the development of secure and efficient OFDM systems that are reliable under hostile jamming, especially the destructive disguised jamming.

If we examine disguised jamming carefully, we can see that the main issue there is the symmetricity between the authorized signal and the jamming interference. Intuitively, to design the corresponding anti-jamming system, the main task is to break the symmetricity between the authorized signal and the jamming interference or make it impossible for the jammer to achieve this symmetricity. For this purpose, encryption or channel coding at the bit level will not really help, since the symmetricity actually appears at the symbol level. That is, instead of using a fixed symbol constellation, we have to introduce secure randomness to our constellation and utilize a dynamic or random constellation scheme, such that the jammer can no longer mimic the authorized user's signal. At the same time, the authorized user does not have to sacrifice too much on the performance, efficiency, and system complexity.

Motivated by the observations above and our previous research on anti-jamming system design [9–12, 14], in this chapter, we present a securely precoded OFDM (SP-OFDM) system for efficient and reliable transmission under disguised jamming. By integrating advanced cryptographic techniques into OFDM transceiver design, we randomize the constellation by introducing shared randomness between the legitimate transmitter and receiver, which breaks the symmetricity between the authorized signal and the jamming interference and hence ensures reliable performance under disguised jamming. More specifically, the basic idea is to randomize the phase of transmitted symbols using the secure PN sequences generated from the Advanced Encryption Standard (AES) algorithm. *A remarkable feature of SP-OFDM is that it achieves strong jamming resistance but has the same high spectral efficiency as the traditional OFDM system. Moreover, the change to the physical layer transceivers is minimal, feasible, and affordable.* The robustness of SP-OFDM under disguised jamming is demonstrated through both theoretic and numerical analyses.

We will analyze the channel capacity of the traditional OFDM and SP-OFDM under hostile jamming using the arbitrarily varying channel (AVC) model. It will be seen that the deterministic code capacity of the traditional OFDM is zero under the worst disguised jamming. At the same time, we will prove that with the secure randomness shared between the authorized transmitter and receiver, the AVC channel corresponding to SP-OFDM is not symmetrizable, and hence SP-OFDM can achieve a positive deterministic code capacity under disguised jamming. Note that the authorized user aims to maximize the capacity, while the jammer aims to minimize the capacity; we will show that the maximin capacity for SP-OFDM under hostile jamming is given by $C = \log\left(1 + \frac{P_S}{P_J+P_N}\right)$ bits per symbol, where P_s denotes the signal power, P_J the jamming power, and P_N the noise power.

We will further evaluate the capacity of SP-OFDM under disguised jamming and analyze the relationship between the constellation size and capacity with respect to the M-ary PSK modulation. PSK is chosen here since it is a constant modulus modulation and every symbol utilizes the peak transmit power. It will be shown that when noise is present, the capacity of SP-OFDM under disguised jamming converges as the constellation size increases. This result justifies the use of practical, finite size constellations in SP-OFDM.

Numerical examples are provided to demonstrate the effectiveness of SP-OFDM under disguised jamming and frequency selective fading. It should be pointed out that *the secure precoding scheme discussed here can also be applied to other MPSK based modulation techniques*, such as EDGE [15].

To this end, we can see that without secure precoding, disguised jamming is the worst jamming that can cause the deterministic code capacity to be zero. An interesting question is when secure precoding is applied, what would be the worst jamming distribution that minimizes the channel capacity of SP-OFDM? In this chapter, we will consider a practical wireless communication scenario, where the transmitting symbol is uniformly distributed in a discrete and finite alphabet, and the jamming interference is subject to finite average and peak power constraints. By formulating the problem as a constrained functional optimization process [16], we can derive the Kuhn-Tucker (KT) conditions that should be satisfied by the worst jamming distribution. We will see that limited by the peak transmit power, the worst jamming for SP-OFDM would be discrete in amplitude with a finite number of mass points. However, due to the inherent secure randomness in SP-OFDM, disguised jamming is no longer the worst jamming for SP-OFDM.

5.2 Secure OFDM System Design Under Disguised Jamming

In this section, we introduce the anti-jamming OFDM system with secure precoding and decoding, named as securely precoded OFDM (SP-OFDM).

5.2.1 *Transmitter Design with Secure Precoding*

The block diagram of the system is shown in Fig. 5.1. Let N_c be the number of subcarriers in the OFDM system and Φ the alphabet of transmitted symbols. For $i = 0, 1, \cdots, N_c - 1$ and $k = 0, 1, \cdots$, let $S_{k,i} \in \Phi$ denote the symbol transmitted on the i-th carrier of the k-th OFDM block. We denote the symbol vector of the k-th OFDM block by $\boldsymbol{S}_k = [S_{k,0}, S_{k,1}, \cdots, S_{k,N_c-1}]^T$. The input data stream is first fed to the channel encoder, mapped to the symbol vector $\boldsymbol{S}_k$, and then fed to the symbol-level secure precoder.

As pointed out in [11, 14, 17, 18], a key enabling factor for reliable communication under disguised jamming is to introduce shared randomness between the

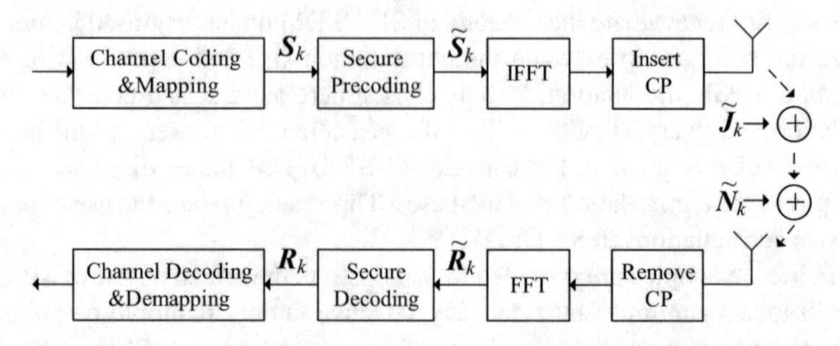

Fig. 5.1 Anti-jamming OFDM design through secure precoding and decoding

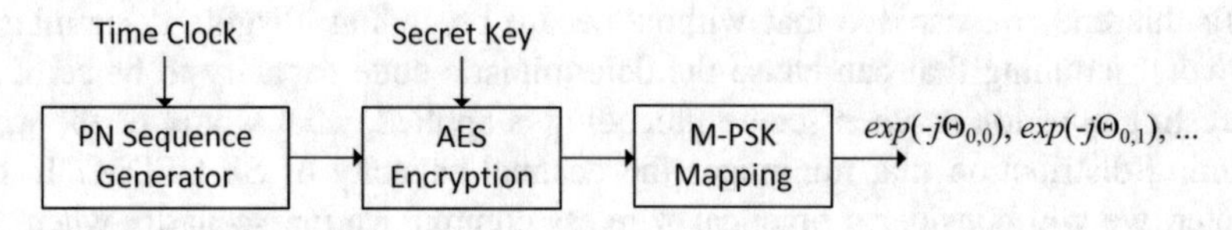

Fig. 5.2 Secure phase-shift generator

transmitter and receiver, such that the symmetricity between the authorized signal and the jamming interference is broken. To maintain full spectral efficiency of the traditional OFDM system, the precoding is performed by multiplying an *invertible* $N_c \times N_c$ precoding matrix $\boldsymbol{P}_k$ to the symbol vector $\boldsymbol{S}_k$, i.e.,

$$\tilde{\boldsymbol{S}}_k = \boldsymbol{P}_k \boldsymbol{S}_k. \tag{5.1}$$

In SP-OFDM, we design the precoding matrix $\boldsymbol{P}_k$ to be a diagonal matrix as

$$\boldsymbol{P}_k = \text{diag}(e^{-j\Theta_{k,0}}, e^{-j\Theta_{k,1}}, \cdots, e^{-j\Theta_{k,N_c-1}}). \tag{5.2}$$

That is, for $i = 0, 1, \cdots, N_c - 1$, and $k = 0, 1, \cdots$, a random phase shift is applied to each transmitted symbol; more specifically, a random phase shift $-\Theta_{k,i}$ is applied to the symbol transmitted on the i-th carrier of the k-th OFDM block. The phase-shift changes randomly and independently across the subcarriers and OFDM blocks and is encrypted so that the jammer has no access to it. More specifically, $\{\Theta_{k,i}\}$ is generated through a secure phase-shift generator as shown in Fig. 5.2. The secure phase-shift generator consists of three parts: (i) a PN sequence generator, (ii) an Advanced Encryption Standard (AES) [19] encryption module, and (iii) an M-PSK mapper.

The *PN sequence generator* generates a pseudorandom sequence, which is then encrypted with AES. The encrypted sequence is further converted to PSK symbols using an M-PSK mapper, where M is a power of 2 and every $\log_2 M$ bits are converted to a PSK symbol. The structure and the initial state of the PN sequence generator are public, so that the transmitter and the receiver can generate identical phase-shift sequences as long as they share the same secret encryption key.

The security, as well as the randomness of the generated phase-shift sequence, is guaranteed by the AES encryption algorithm [19], for which the secret encryption key is only shared between the authorized transmitter and receiver. As a result, the phase-shift sequence is random and inaccessible for the jammer.

The resulted symbol vector from the secure precoding, $\tilde{\boldsymbol{S}}_k$, is then used to generate the OFDM signals through IFFT and transmitted to the receiver after inserting the cyclic prefix (CP), which eliminates the intersymbol interference (ISI) introduced by the multipath channels.

5.2.2 Receiver Design with Secure Decoding

We consider an additive white Gaussian noise (AWGN) channel under hostile jamming. The transmitted OFDM signal is subject to an AWGN term, denoted by $\tilde{\boldsymbol{N}}_k$, and an additive jamming interference $\tilde{\boldsymbol{J}}_k$. Let $\tilde{\boldsymbol{N}}_k = [\tilde{N}_{k,0}, \tilde{N}_{k,1}, \cdots, \tilde{N}_{k,N_c-1}]$. We assume that the components in $\tilde{\boldsymbol{N}}_k$ are i.i.d. circularly symmetric complex Gaussian random variables. Note that in OFDM, the original symbol vectors $\{\boldsymbol{S}_k\}$ are considered to be in the frequency domain for $k = 0, 1, \cdots$; we also express the noise term $\tilde{\boldsymbol{N}}_k$ and the jamming interference $\tilde{\boldsymbol{J}}_k$ at the receiver input (please refer to Fig. 5.1) in terms of their frequency components as

$$\tilde{\boldsymbol{N}}_k = \boldsymbol{F}\bar{\boldsymbol{N}}_k, \text{ with } \bar{\boldsymbol{N}}_k = [\bar{N}_{k,0}, \bar{N}_{k,1}, \cdots, \bar{N}_{k,N_c-1}]^T, \tag{5.3}$$

$$\tilde{\boldsymbol{J}}_k = \boldsymbol{F}\boldsymbol{J}_k, \text{ with } \boldsymbol{J}_k = [J_{k,0}, J_{k,1}, \cdots, J_{k,N_c-1}]^T, \tag{5.4}$$

where $\bar{N}_{k,i}, J_{k,i} \in \mathbb{C}, i = 0, 1, \cdots, N_c - 1$, denote the noise and the jamming interference on the i-th subcarrier of the k-th OFDM block, respectively; $\boldsymbol{F}$ is the IFFT unitary matrix with $[\boldsymbol{F}]_{m,n} = \frac{1}{\sqrt{N_c}} e^{j2\pi mn/N_c}$, and $\mathbb{C}$ is the complex plane.

On the receiver side, after removing the CP and performing OFDM demodulation (FFT) to the received signal, a symbol vector $\tilde{\boldsymbol{R}}_k = [\tilde{R}_{k,0}, \tilde{R}_{k,1}, \cdots, \tilde{R}_{k,N_c-1}]^T$ is obtained for the k-th transmitted OFDM block. That is,

$$\tilde{\boldsymbol{R}}_k = \boldsymbol{P}_k\boldsymbol{S}_k + \boldsymbol{J}_k + \bar{\boldsymbol{N}}_k. \tag{5.5}$$

Note that the noise vector $\tilde{\boldsymbol{N}}_k$ has zero mean and covariance matrix $\mathbb{E}\{\tilde{\boldsymbol{N}}_k^H\tilde{\boldsymbol{N}}_k\} = \sigma^2\boldsymbol{I}$; it follows that the corresponding covariance matrix of $\bar{\boldsymbol{N}}_k$ is $\mathbb{E}\{\bar{\boldsymbol{N}}_k^H\bar{\boldsymbol{N}}_k\} = \sigma^2\boldsymbol{I}$, since $\bar{\boldsymbol{N}}_k = F^H\tilde{\boldsymbol{N}}_k$ and F is a unitary matrix. Moreover, note that (i) for any circularly symmetric complex Gaussian random variable N, $e^{j\theta}N$ and N have the same distribution for any angle θ [20, p66]; (ii) linear combination of independent circularly symmetric complex Gaussian random variables is still circularly symmetric complex Gaussian. We can see that the components of noise $\bar{\boldsymbol{N}}_k$ are i.i.d. circularly symmetric complex Gaussian random variables with zero mean and variance σ^2.

The secure decoding module multiplies the inverse matrix of $\boldsymbol{P}_k$ to $\tilde{\boldsymbol{R}}_k$, which results in the symbol vector

$$\boldsymbol{R}_k = \boldsymbol{S}_k + \boldsymbol{P}_k^{-1}\boldsymbol{J}_k + \boldsymbol{P}_k^{-1}\bar{\boldsymbol{N}}_k, \tag{5.6}$$

where $\boldsymbol{R}_k = [R_{k,0}, R_{k,1}, \cdots, R_{k,N_c-1}]^T$, with

$$R_{k,i} = S_{k,i} + e^{j\Theta_{k,i}} J_{k,i} + N_{k,i}, \tag{5.7}$$

where $N_{k,i} = e^{j\Theta_{k,i}}\bar{N}_{k,i}$ and $\Theta_{k,i}$ is uniformly distributed over $\{\frac{2\pi l}{M} \mid l = 0, 1, \cdots, M-1\}$. Again, the phase shift to the complex Gaussian noise $\bar{N}_{k,i}$ will not change its distribution. That is, $N_{k,i}$ is still a circular symmetric complex Gaussian random variable of zero mean and variance σ^2. For notation simplicity, from now on, we replace the double indices k, i of each symbol with a single index n, where $n = kN_c + i$.

Taking the delay in communication system into consideration, we assume that the authorized user and the jammer do not have pre-knowledge on the transmitted sequence of each other. We do not assume any a priori information on the jamming signal, except a finite average power constraint of P_J, i.e., $\mathbb{E}\{|J_n|^2\} \leq P_J$, for any possible n.

5.2.3 PN Sequence Synchronization Between the Secure Precoder and Decoder

A key issue in the secure precoding and decoding is the PN sequence synchronization. We ensure the synchronization between the PN sequences generated by the transmitter and the receiver using the following method: each party is equipped with a global time clock, and the PN sequence generators of the two parties are reinitialized periodically at fixed intervals. The new state for reinitialization, for example, can be the elapsed time after a specific reference epoch in seconds for the time being, which is public. As the initial state changes with each reinitialization, no repeated PN sequence will be generated.

At the beginning of reception, the PN sequences generated by the two parties may not be perfectly synchronized because of the mismatch between the time clocks and the possible delays in the transmission and reception. That is, the receiver may be unaware of the indices of the symbols at first.

This problem can be solved by inserting pilot symbols into the transmitted signal. Considering the random phase shifts introduced in the secure precoding, without loss of generality, we can assume the pilot symbol vector is $\boldsymbol{s}_p[1, 1, \cdots, 1]^T_{1\times m}$, where m is the length of pilot sequence. Suppose the received training sequence is denoted as $[\tilde{R}_{n_0}, \tilde{R}_{n_0+1}, \ldots \tilde{R}_{n_0+m-1}]^T$, where the starting index n_0 is unknown to the receiver, i.e.,

$$\tilde{R}_{n_0+l} = \boldsymbol{s}_p e^{-j\Theta_{n_0+l}} + J_{n_0+l} + N_{n_0+l}, \quad l = 0, 1, \cdots, m-1. \tag{5.8}$$

To estimate the value of n_0, the receiver calculates its correlation with the known phase-shifting sequence $e^{j\Theta_{n_1}}, e^{j\Theta_{n_1+1}}, \ldots, e^{j\Theta_{n_1+m-1}}$, i.e.,

$$\frac{1}{m}\sum_{l=0}^{m-1} \tilde{R}_{n_0+l} e^{j\Theta_{n_1+l}} = \frac{1}{m}\sum_{l=0}^{m-1} \left[s_p e^{-j(\Theta_{n_1+l}-\Theta_{n_0+l})} + J_{n_0+l} e^{j\Theta_{n_1+l}} + N_{n_0+l} \right]. \quad (5.9)$$

The receiver iterates n_1 over the candidate set of n_0, that is, all the possible values of n_0. Note that for $l_1 \neq l_2$, Θ_{l_1} and Θ_{l_2} are independent, and J_{l_1} and Θ_{l_2} are also independent, so we can get the mean and variance of the correlation coefficient as

$$\mathbb{E}\left[\frac{1}{m}\sum_{l=0}^{m-1} \tilde{R}_{n_0+l} e^{j\Theta_{n_1+l}} \right] = \begin{cases} 0, & n_0 \neq n_1 \\ s_p, & n_0 = n_1 \end{cases}, \quad (5.10)$$

$$\mathbb{D}\left[\frac{1}{m}\sum_{l=0}^{m-1} \tilde{R}_{n_0+l} e^{j\Theta_{n_1+l}} \right] \leq \begin{cases} \frac{|s_p|^2+P_J+\sigma_n^2}{m}, & n_0 \neq n_1 \\ \frac{P_J+\sigma_n^2}{m}, & n_0 = n_1 \end{cases}. \quad (5.11)$$

The value of n_0 can then be estimated through

$$\hat{n}_0 = \arg\max_{n_1} \left| \frac{1}{m}\sum_{l=0}^{m-1} \tilde{R}_{n_0+l} e^{j\Theta_{n_1+l}} \right|. \quad (5.12)$$

Note that from the design of the PN generator in Sect. 5.2.1, the receiver is able to have a rough estimation of n_0 from the time clock, so the candidate set of n_0 is finite. Under limited jamming power, the variance of the estimated correlation coefficient can be arbitrarily small as $m \to \infty$. Thus from the Chebyshev inequality [21, Theorem 5.11], for any $n_1 \neq n_0$, the probability

$$\Pr\left\{ \left| \frac{1}{m}\sum_{l=0}^{m-1} \tilde{R}_{n_0+l} e^{j\Theta_{n_1+l}} \right| \geq \left| \frac{1}{m}\sum_{l=0}^{m-1} \tilde{R}_{n_0+l} e^{j\Theta_{n_0+l}} \right| \right\} \to 0 \text{ as } m \to \infty, \quad (5.13)$$

which indicates we are able to get an accurate estimate of n_0.

Basing on the analysis above, in the following, we assume the PN sequences have been perfectly synchronized between the two parties. For notation simplicity, we further discard the indices of symbol and rewrite the equivalent channel model as

$$R = S + e^{j\Theta} J + N, \quad (5.14)$$

where $S \in \Phi$, $J \in \mathbb{C}$, $N \sim \mathcal{CN}(0, \sigma^2)$, Θ is uniformly distributed over $\{\frac{2\pi l}{M} \mid l = 0, 1, \ldots, M-1\}$ and $\mathcal{CN}(\boldsymbol{\mu}, \boldsymbol{\Sigma})$ denotes a circularly symmetric complex Gaussian distribution with mean $\boldsymbol{\mu}$ and variance $\boldsymbol{\Sigma}$. We would like to point out that this model is used for system capacity evaluation under disguised jamming. The system performance with non-ideal carrier synchronization or channel estimation under fading channels will be demonstrated in Example 2 of Sect. 5.4, where the error in carrier synchronization or channel estimation is modeled as the random channel gain under a Rician model.

5.3 Symmetricity and Capacity Analysis using the AVC Model

In this section, first, we will show that for the SP-OFDM system, the equivalent arbitrarily varying channel (AVC) model is nonsymmetrizable under disguised jamming. We will further discuss the capacity of SP-OFDM under hostile jamming.

5.3.1 AVC Symmetricity Analysis

Recall that the arbitrarily varying channel (AVC) model [18] characterizes the communication channels with unknown states which may vary in arbitrary manners across time. For the jamming channel (5.14) of interest,

$$R = S + e^{j\Theta} J + N,$$

the jamming symbol J can be viewed as the state of the channel under consideration. The channel capacity of AVC evaluates the data rate of the channel under the most adverse jamming interference among all the possibilities [22]. Note that unlike the jamming free model where the channel noise sequence is independent of the authorized signal and is i.i.d., the AVC model considers the possible correlation between the authorized signal and the jamming, as well as the possible temporal correlation among the jamming symbols, which may cause much worse damages to the communication.

To prove the effectiveness of SP-OFDM under disguised jamming, we need to introduce some basic concepts and properties of the AVC model. First we revisit the definition of symmetrizable AVC channel.

Definition 5.1 ([22, 23]) Let $W(\boldsymbol{r} \mid \boldsymbol{s}, \boldsymbol{x})$ denote the conditional PDF of the received signal R given the transmitted symbol $\boldsymbol{s} \in \Phi$ and the jamming symbol $\boldsymbol{x} \in \mathbb{C}$. The AVC channel (5.14) is symmetrizable if for some auxiliary channel $\pi : \Phi \to \mathbb{C}$, we have

$$\int_{\mathbb{C}} W(\boldsymbol{r} \mid \boldsymbol{s}, \boldsymbol{x}) \mathrm{d}F_{\pi}(\boldsymbol{x}|\boldsymbol{s}') = \int_{\mathbb{C}} W(\boldsymbol{r} \mid \boldsymbol{s}', \boldsymbol{x}) \mathrm{d}F_{\pi}(\boldsymbol{x}|\boldsymbol{s}), \quad \forall \boldsymbol{s}, \boldsymbol{s}' \in \Phi, \boldsymbol{r} \in \mathbb{C}, \tag{5.15}$$

where $F_\pi(\cdot|\cdot)$ is the probability measure of the output of channel π given the input, i.e., the conditional CDF

$$F_\pi(\boldsymbol{x}|\boldsymbol{s}) = \Pr\{Re(\pi(\boldsymbol{s})) \le Re(\boldsymbol{x}), Im(\pi(\boldsymbol{s})) \le Im(\boldsymbol{x})\}, \quad \boldsymbol{x} \in \mathbb{C}, \ \boldsymbol{s} \in \Phi, \tag{5.16}$$

where $\pi(\boldsymbol{s})$ denotes the output of channel π given input symbol $\boldsymbol{s}$.

We denote the set of all the auxiliary channels, π's, that can symmetrize channel (5.14) by Π, that is,

$$\Pi = \left\{\pi \mid \text{Eq. (5.15) is satisfied w.r.t. } \pi \text{ for any } \boldsymbol{s}, \boldsymbol{s}' \in \Phi \text{ and } \boldsymbol{r} \in \mathbb{C}\right\}. \tag{5.17}$$

With the average jamming power constraint considered, we further introduce the definition of l-symmetrizable channel.

Definition 5.2 ([23]) The AVC channel (5.14) is called l-symmetrizable under average jamming power constraint if and only if (iff) there exists a $\pi \in \Pi$ such that

$$\int_{\mathbb{C}} |\boldsymbol{x}|^2 \mathrm{d}F_\pi(\boldsymbol{x}|\boldsymbol{s}) < \infty, \quad \forall \boldsymbol{s} \in \Phi. \tag{5.18}$$

In [23], it was shown that reliable communication can be achieved as long as the AVC channel is not l-symmetrizable.

Lemma 5.1 ([23, Corollary 2]) *The deterministic code capacity of AVC channel* (5.14) *is positive under any hostile jamming with finite average power constraint iff the AVC is not l-symmetrizable. Furthermore, given a specific average jamming power constraint P_J, the channel capacity C in this case equals*

$$\begin{aligned} C = \max_{\mathcal{P}_S} \min_{F_J} I(S, R), \\ s.t. \ \int_{\mathbb{C}} |\boldsymbol{x}|^2 dF_J(\boldsymbol{x}) \le P_J, \end{aligned} \tag{5.19}$$

where $I(S, R)$ denotes the mutual information (MI) between the R and S in (5.14) *and $\mathcal{P}_S$ denotes the probability distribution of S over Φ and $F_J(\cdot)$ the CDF of J.*

First, we show that the traditional OFDM system is l-symmetrizable under disguised jamming.

Theorem 5.1 *The traditional OFDM system is l-symmetrizable. Therefore, the deterministic code capacity is zero under disguised jamming with finite average jamming power.*

Proof The AVC model of the traditional OFDM system is

$$R = S + J + N. \tag{5.20}$$

We will show that when S and J have the same constellation Φ, hence the same finite average power, the AVC channel is l-symmetrizable. It follows from (5.20) that

$$W(\boldsymbol{r} \mid \boldsymbol{s}, \boldsymbol{s}') = W(\boldsymbol{r} \mid \boldsymbol{s}', \boldsymbol{s}), \quad \forall \boldsymbol{s}, \boldsymbol{s}' \in \Phi, \ \boldsymbol{r} \in \mathbb{C}. \tag{5.21}$$

Since Φ has finite average power, the average power constraint (5.18) is satisfied by the disguised jamming. Hence, by selecting channel π given input $\boldsymbol{x}$ as

$$\pi(\boldsymbol{x}) = \boldsymbol{x}, \tag{5.22}$$

channel (5.20) is l-symmetrizable by π. From Lemma 5.1, a necessary condition for a positive AVC deterministic code capacity is that the channel is not l-symmetrizable. So the traditional OFDM system has zero deterministic code capacity under disguised jamming with finite average jamming power. ∎

Next, we show that with the secure precoding, it is impossible to l-symmetrize the AVC channel (5.14) corresponding to the SP-OFDM system.

Theorem 5.2 *The AVC channel corresponding to SP-OFDM is not l-symmetrizable.*

Proof We prove this result by contradiction. Suppose that there exists a channel $\pi \in \Pi$ such that the AVC channel is l-symmetrizable. Denote the output of channel π given input $\boldsymbol{x}$ by $\pi(\boldsymbol{x})$, and define the corresponding AVC channel output for inputs $\boldsymbol{s}$ and $\boldsymbol{s}'$ as

$$\hat{R}(\boldsymbol{s}, \boldsymbol{s}') = \boldsymbol{s} + \pi(\boldsymbol{s}')e^{j\Theta} + N, \tag{5.23}$$

where $\hat{R}(\boldsymbol{s}, \boldsymbol{s}')$ denotes the channel output. Following (5.15), $\hat{R}(\boldsymbol{s}, \boldsymbol{s}')$ and $\hat{R}(\boldsymbol{s}', \boldsymbol{s})$ have the same distribution. Let $\varphi_X(\omega_1, \omega_2)$ denote the characteristic function (CF) of a complex random variable X. So we have

$$\varphi_{\hat{R}(\boldsymbol{s},\boldsymbol{s}')}(\omega_1, \omega_2) \equiv \varphi_{\hat{R}(\boldsymbol{s}',\boldsymbol{s})}(\omega_1, \omega_2), \tag{5.24}$$

and

$$\varphi_{\hat{R}(\boldsymbol{s},\boldsymbol{s}')}(\omega_1, \omega_2) = \varphi_{[\boldsymbol{s}+\pi(\boldsymbol{s}')e^{j\Theta}]}(\omega_1, \omega_2)\, \varphi_N(\omega_1, \omega_2), \tag{5.25}$$

where, for the complex Gaussian noise N, we have

$$\varphi_N(\omega_1, \omega_2) = e^{-\frac{\sigma^2}{4}(w_1^2+w_2^2)}, \quad \omega_1, \omega_2 \in (-\infty, +\infty), \tag{5.26}$$

which is nonzero over $\mathbb{R}^2$. Thus by eliminating the characteristic functions of the Gaussian noises on both sides of Eq. (5.24), we have

$$\varphi_{[\boldsymbol{s}+\pi(\boldsymbol{s}')e^{j\Theta}]}(\omega_1, \omega_2) = \varphi_{[\boldsymbol{s}'+\pi(\boldsymbol{s})e^{j\Theta}]}(\omega_1, \omega_2). \tag{5.27}$$

for $\omega_1, \omega_2 \in (-\infty, +\infty)$. Let $\boldsymbol{s} = s_1 + js_2$, we can then express $\varphi_{[\boldsymbol{s}+\pi(\boldsymbol{s}')e^{j\Theta}]}(\omega_1, \omega_2)$ as

$$\varphi_{[\boldsymbol{s}+\pi(\boldsymbol{s}')e^{j\Theta}]}(\omega_1, \omega_2) = e^{js_1\omega_1 + js_2\omega_2}\varphi_{[\pi(\boldsymbol{s}')e^{j\Theta}]}(\omega_1, \omega_2), \tag{5.28}$$

and

$$\begin{aligned}\varphi_{[\pi(\boldsymbol{s}')e^{j\Theta}]}(\omega_1, \omega_2) &= \mathbb{E}\{e^{j\omega_1 Re(\pi(\boldsymbol{s}')e^{j\Theta}) + j\omega_2 Im(\pi(\boldsymbol{s}')e^{j\Theta})}\} \\ &= \int_{\mathbb{C}} \mathbb{E}\{e^{j\omega_1 Re(\boldsymbol{x}e^{j\Theta}) + j\omega_2 Im(\boldsymbol{x}e^{j\Theta})}\} \mathrm{d}F_\pi(\boldsymbol{x}|\boldsymbol{s}'). \end{aligned} \tag{5.29}$$

Recall that under the secure precoding scheme, Θ is uniformly distributed over $\{\frac{2\pi l}{M} \mid l = 0, 1, \ldots, M-1\}$, where M is a power of 2. We have

$$\begin{aligned}&\mathbb{E}\{e^{j\omega_1 Re(\boldsymbol{x}e^{j\Theta}) + j\omega_2 Im(\boldsymbol{x}e^{j\Theta})}\} \\ &= \frac{1}{M}\sum_{i=0}^{M-1} e^{j\omega_1|\boldsymbol{x}|\cos(\frac{2\pi i}{M} + \arg(\boldsymbol{x})) + j\omega_2|\boldsymbol{x}|\sin(\frac{2\pi i}{M} + \arg(\boldsymbol{x}))} \\ &= \frac{1}{M}\left(\sum_{i=0}^{M/2-1} + \sum_{i=M/2}^{M-1}\right) e^{j\omega_1|\boldsymbol{x}|\cos(\frac{2\pi i}{M} + \arg(\boldsymbol{x})) + j\omega_2|\boldsymbol{x}|\sin(\frac{2\pi i}{M} + \arg(\boldsymbol{x}))} \\ &= \frac{2}{M}\sum_{i=0}^{M/2-1} \cos\left[\omega_1|\boldsymbol{x}|\cos\left(\frac{2\pi i}{M} + \arg(\boldsymbol{x})\right) + \omega_2|\boldsymbol{x}|\sin\left(\frac{2\pi i}{M} + \arg(\boldsymbol{x})\right)\right],\end{aligned} \tag{5.30}$$

which is of real value for $\omega_1, \omega_2 \in (-\infty, +\infty)$. So $\varphi_{[\pi(\boldsymbol{s}')e^{j\Theta}]}(\omega_1, \omega_2)$ and $\varphi_{[\pi(\boldsymbol{s})e^{j\Theta}]}(\omega_1, \omega_2)$ are also real-valued over $\mathbb{R}^2$. For $\boldsymbol{s} \neq \boldsymbol{s}'$ and $\boldsymbol{s}' = s_1' + js_2'$, (5.27) can be expressed as

$$\varphi_{[\pi(\boldsymbol{s}')e^{j\Theta}]}(\omega_1, \omega_2) = e^{j[(s_1 - s_1')\omega_1 + (s_2 - s_2')\omega_2]}\varphi_{[\pi(\boldsymbol{s})e^{j\Theta}]}(\omega_1, \omega_2), \tag{5.31}$$

where $e^{j[(s_1 - s_1')\omega_1 + (s_2 - s_2')\omega_2]}$ has nonzero imaginary part for $(s_1 - s_1')\omega_1 + (s_2 - s_2')\omega_2 \neq n\pi$, $n \in \mathbb{Z}$. Without loss of generality, we assume $s_1 \neq s_1'$. From (5.30) and (5.31), we have

$$\varphi_{[\pi(\boldsymbol{s})e^{j\Theta}]}(\omega_1, \omega_2) = 0, \text{ for } \omega_1 + \frac{s_2 - s_2'}{s_1 - s_1'}\omega_2 \neq \frac{n\pi}{s_1 - s_1'}, \forall n \in \mathbb{Z}. \tag{5.32}$$

On the other hand, the characteristic function of an RV should be uniformly continuous in the real domain [21, Theorem 15.21]. So for any fixed $\omega_2 \in (-\infty, \infty)$, we should have

$$\varphi_{[\pi(s)e^{j\Theta}]}(\frac{n\pi - (s_2 - s_2')\omega_2}{s_1 - s_1'}, \omega_2) = \lim_{\omega_1 \to \frac{n\pi - (s_2 - s_2')\omega_2}{s_1 - s_1'}} \varphi_{[\pi(s)e^{j\Theta}]}(\omega_1, \omega_2), \ \forall n \in \mathbb{Z}. \tag{5.33}$$

For $\omega_1 \in (\frac{(n-1)\pi - (s_2 - s_2')\omega_2}{s_1 - s_1'}, \frac{n\pi - (s_2 - s_2')\omega_2}{s_1 - s_1'}) \cup (\frac{n\pi - (s_2 - s_2')\omega_2}{s_1 - s_1'}, \frac{(n+1)\pi - (s_2 - s_2')\omega_2}{s_1 - s_1'})$,

$$\varphi_{[\pi(s)e^{j\Theta}]}(\omega_1, \omega_2) \equiv 0, \tag{5.34}$$

so

$$\varphi_{[\pi(s)e^{j\Theta}]}(\frac{n\pi - (s_2 - s_2')\omega_2}{s_1 - s_1'}, \omega_2) = 0, \ \forall n \in \mathbb{Z}. \tag{5.35}$$

Combining (5.32) and (5.35), we have

$$\varphi_{[\pi(s)e^{j\Theta}]}(\omega_1, \omega_2) = 0, \ \ \forall \omega_1, \omega_2 \in (-\infty, \infty). \tag{5.36}$$

However, (5.36) cannot be a valid characteristic function for any RV. Therefore, the auxiliary channel π does not exist, and Π is empty. Hence, the AVC channel is not l-symmerizable. ■

Following Lemma 5.1, the result in Theorem 5.2 implies that SP-OFDM will always have positive deterministic code capacity under any hostile jamming with finite average power. The next subsection is focused on how to calculate the channel capacity of SP-OFDM under hostile jamming.

5.3.2 Capacity Analysis

From Lemma 5.1, the capacity of channel $R = S + e^{j\Theta} J + N$ is given by

$$C = \max_{\mathcal{P}_S} \min_{F_J} I(S, R),$$
$$s.t. \ \int_{\mathbb{C}} |\boldsymbol{x}|^2 \mathrm{d}F_J(\boldsymbol{x}) \leq P_J.$$

It is hard to obtain a closed form solution of the channel capacity for a general discrete transmission alphabet Φ. However, if we relax the distribution of the transmitted symbol S from the discrete set Φ to the entire complex plane $\mathbb{C}$ under an average power constraint, we are able to obtain the following result on channel capacity.

Theorem 5.3 *The deterministic code capacity of SP-OFDM is positive under any hostile jamming. More specifically, let the alphabet $\Phi = \mathbb{C}$ and the average power of S being upper bounded by P_S, then the maximin channel capacity in* (5.19) *under* *average jamming power constraint P_J and noise power $P_N = \sigma^2$ is*

$$C = \log\left(1 + \frac{P_S}{P_J + P_N}\right). \tag{5.37}$$

The capacity is achieved when the input $S \sim \mathcal{CN}(0, P_S)$ and the jamming $J \sim \mathcal{CN}(0, P_J)$.

To prove Theorem 5.3, we need the following lemma [23, Lemma 4].

Lemma 5.2 *Mutual information $I(S, R)$ is concave with respect to the input distribution $F_S(\cdot)$ and convex with respect to the jamming distribution $F_J(\cdot)$.*

Proof (Proof of Theorem 5.3*)* First, following Lemma 5.1 and Theorem 5.2, we can get that deterministic code capacity of SP-OFDM is positive under any hostile jamming.

Second, we will evaluate the channel capacity of SP-OFDM under hostile jamming. When the support of S is $\Phi = \mathbb{C}$, the whole complex plane, following Lemma 5.1, the channel capacity in (5.19) equals:

$$C = \max_{F_S} \min_{F_J} I(S, R), \tag{5.38}$$

$$s.t. \ \int_{\mathbb{C}} |\boldsymbol{x}|^2 \mathrm{d}F_S(\boldsymbol{x}) \leq P_S, \tag{5.39}$$

$$\int_{\mathbb{C}} |\boldsymbol{x}|^2 \mathrm{d}F_J(\boldsymbol{x}) \leq P_J, \tag{5.40}$$

where $F_S(\cdot)$ denotes the CDF function of S defined on $\mathbb{C}$ and (5.39) and (5.40) denote the average power constraints on the input and the jamming, respectively.

We denote the $I(S, R)$ w.r.t the input distribution $F_S(\cdot)$ and the jamming distribution $F_J(\cdot)$ by $\phi(F_S, F_J)$. Following Lemma 5.2, $\phi(F_S, F_J)$ is concave w.r.t. $F_S(\cdot)$ and convex w.r.t. $F_J(\cdot)$. As shown in [24], if we can find the input distribution F_S^* and the jamming distribution F_J^* such that

$$\phi(F_S, F_J^*) \leq \phi(F_S^*, F_J^*) \leq \phi(F_S^*, F_J), \tag{5.41}$$

for any F_S and F_J satisfying the average power constraints (5.39) and (5.40), respectively, then

$$\phi(F_S^*, F_J^*) = C. \tag{5.42}$$

That is, the pair (F_S^*, F_J^*) is the saddle point of the maximin problem in Eq. (5.38) [25].

Assume the jamming interference is circularly symmetric complex Gaussian with average power P_J, that is, $F_J^* = \mathcal{CN}(0, P_J)$. Note that the phase shift would not change the distribution of a complex Gaussian RV and the fact that the jamming J and the noise N are independent. The jammed channel in this case is equivalent to a complex AWGN channel with noise power $P_J + P_N$, where the capacity achieving input distribution is also a complex Gaussian with power P_S, that is, $F_S^* = \mathcal{CN}(0, P_S)$. It follows that for any input distribution F_S satisfying the power constraint P_S,

$$\phi(F_S, \mathcal{CN}(0, P_J)) \leq \phi(\mathcal{CN}(0, P_S), \mathcal{CN}(0, P_J)). \tag{5.43}$$

On the other hand, when the input distribution is $F_S^* = \mathcal{CN}(0, P_S)$, the worst noise in terms of capacity for Gaussian input is Gaussian [26]. Since $e^{j\Theta} J + N$ is complex Gaussian with power $P_J + P_N$ if $F_J^* = \mathcal{CN}(0, P_J)$, then for any jamming distribution F_J satisfying the power constraint P_J,

$$\phi(\mathcal{CN}(0, P_S), \mathcal{CN}(0, P_J)) \leq \phi(\mathcal{CN}(0, P_S), F_J). \tag{5.44}$$

So the saddle point (F_S^*, F_J^*) is achieved at $(\mathcal{CN}(0, P_S), \mathcal{CN}(0, P_J))$, where the corresponding channel capacity is

$$C = \log\left(1 + \frac{P_S}{P_J + P_N}\right), \tag{5.45}$$

which completes the proof. ■

5.4 Performance of SP-OFDM Under Disguised Jamming

In this section, we evaluate the performance of SP-OFDM under disguised jamming attacks through simulation examples.

Example 1 (System performance under disguised jamming in AWGN channels) In this example, we analyze the bit error rates (BERs) of SP-OFDM under disguised jamming in AWGN channels. We use the low-density parity check (LDPC) codes for channel coding and adopt the parity check matrices from the DVB-S.2 standard [27]. The coded bits are mapped into QPSK symbols. The random phase shifts in the secure precoding are approximated as i.i.d. continuous RVs uniformly distributed over $[0, 2\pi)$. It can be seen that such an approximation has negligible difference on BER performance compared with a sufficiently large M. The jammer randomly selects one of the code words in the LDPC codebook and sends it to the receiver after the mapping and modulation. On the receiver side, we use a soft decoder for the LDPC codes, where the belief propagation (BP) algorithm [28] is employed. The likelihood information in the BP algorithm is calculated using the likelihood function of a general Gaussian channel, where the noise power is set to $1 + \sigma^2$ considering the existence of jamming signal and σ^2 is the noise power. That is, the signal to jamming power ratio (SJR) is set to be 0 dB. It should be noted that for more complicated jamming distribution or mapping schemes, customized

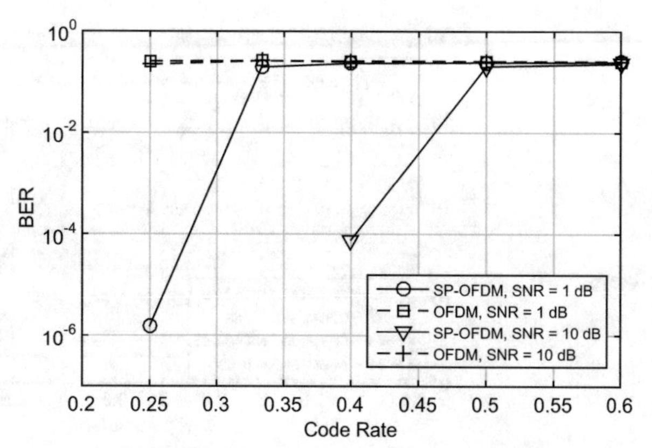

Fig. 5.3 BER performance comparison under disguised jamming in AWGN channels: SP-OFDM versus the traditional OFDM system, signal to jamming power ratio (SJR) = 0 dB

likelihood functions basing on the jamming distribution will be needed for the optimal performance. Figure 5.3 compares the BERs of the communication system studied with and without the secure precoding under different code rates and SNRs. It can be observed that (i) under disguised jamming, in the traditional OFDM system, the BER cannot really be reduced by decreasing the code rate or the noise power, which indicates that without appropriate anti-jamming procedures, we are not able to communicate reliably under disguised jamming; (ii) with the SP-OFDM scheme, when the code rates are below certain thresholds, the BERs can be significantly reduced with the decrease of code rates using the secure precoding. This demonstrates that the SP-OFDM system can achieve a positive deterministic code capacity under disguised jamming.

Example 2 (System performance under disguised jamming in Rician channels) In this example, we verify the effectiveness of SP-OFDM in fading channels. We consider a Rician channel, where the multipath interference is introduced and a strong line of sight (LOS) signal exists [29]. The fading effect is slow enough so that the channel remains unchanged for one OFDM symbol duration. In the simulation, we set the power of the direct path of Rician channel to be 1 and vary the K parameter, which is the ratio between the power of the direct path and that of the scattered path. Figure 5.4 shows the BERs for LDPC code rate $1/3$ under disguised jamming. It can be observed that SP-OFDM is still effective under the fading channel with a sufficiently large K parameter. For a small K parameter, i.e., when the fading is severe, channel estimation and equalization will be needed to guarantee reliable communications.

Example 3 (Effect of phase-shift resolution $\frac{2\pi}{M}$) Recall that in the secure precoding, the secure phase shift is uniformly distributed over $\{\frac{2\pi l}{M} \mid l = 0, 1, \cdots, M-1\}$. In this example, we study the effect of M on the channel capacity of (5.14) under disguised jamming. We consider an 8-PSK mapping and evaluate the channel capacity for different M. The transmitted symbol S and jamming symbol J are

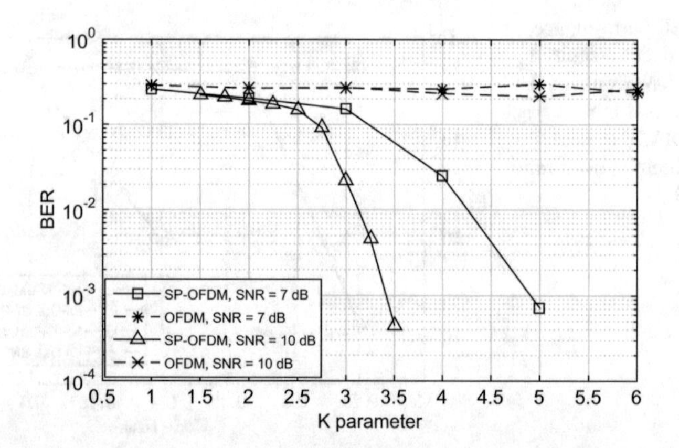

Fig. 5.4 BER performance comparison under disguised jamming in Rician channels: code rate = 1/3, SJR = 0 dB. Here the K parameter refers to the power ratio between the direct path and the scattered path

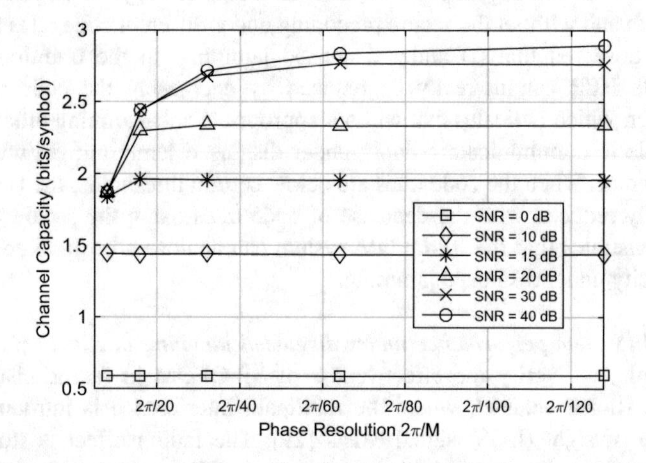

Fig. 5.5 Channel capacity of SP-OFDM under disguised jamming for different values of phase-shift resolution $2\pi/M$, SJR = 0 dB

modeled as uniformly distributed over the 8-PSK constellation. Figure 5.5 shows the channel capacity under different noise powers. It can be observed that the effect of M is related to the SNR level. For a lower SNR level or a sufficiently large M, the change on channel capacity is negligible as M increases. But for reasonably high SNRs, the channel capacity becomes an increasing function with respect to M.

Example 4 (Convergence of channel capacity with the constellation size M_Φ*)* In this example, we evaluate the channel capacity of SP-OFDM using M_Φ-PSK modulation for different constellation size M_Φ. We set $M = 64$ throughout the numerical computation and the power of legitimate sender and hostile jammer are

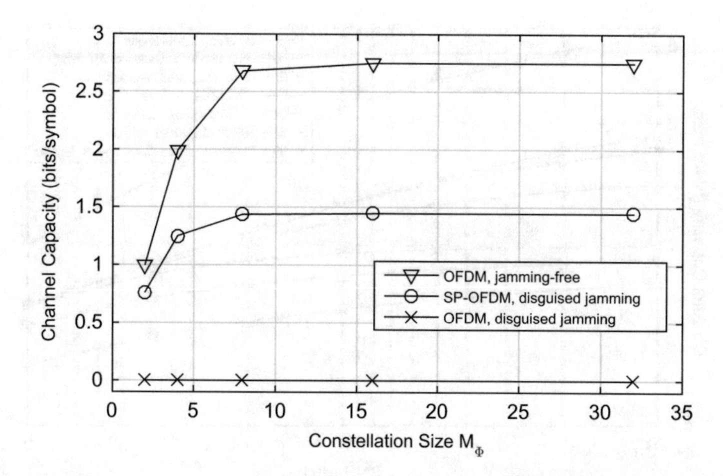

Fig. 5.6 Channel capacity under disguised jamming versus constellation size M_Φ for SNR = 10 dB and SJR = 0 dB

both normalized to 1. Figure 5.6 compares the channel capacity of SP-OFDM versus that of the traditional OFDM under disguised jamming. The capacity of OFDM in the jamming-free case is also provided as a benchmark. Here, we set SNR to be 10 dB. Since the legitimate signal and the additive jamming interference are symmetric in the traditional OFDM system, the deterministic code capacity is 0. The deterministic code capacity of SP-OFDM is nonzero. We can also observe that for the noise power considered, the channel capacity of SP-OFDM under disguised jamming converges as $M_\Phi \to \infty$.

Example 5 (Effect of SNR) In this example, we study the effect of the SNR level on the channel capacity of SP-OFDM. Figure 5.7 shows the channel capacity versus SNR for different modulation schemes. For comparison, the channel capacities of Gaussian input under Gaussian jamming (the saddle-point solution in Theorem 5.3), Gaussian input without jamming, and BPSK modulation under disguised jamming using a nearest neighbor (NN) decoder are also plotted. We can see that (1) for low SNR levels, the M_Φ-PSK ($M_\Phi > 2$) modulation achieves similar capacities under disguised jamming with that shown by the saddle-point solution. This is because for large noise power, the Gaussian mixture distribution can be effectively approximated by a single Gaussian distribution [30], which results in equivalent differential entropies of R for M_Φ-PSK modulation and Gaussian input; (2) for higher SNR levels, the M_Φ-PSK ($M_\Phi > 2$) modulation achieves much higher capacities under disguised jamming than that shown in the saddle-point solution; (3) for BPSK under disguised jamming, the capacity loss after applying NN decoder is less significant on higher SNR levels. This indicates that, for BPSK modulation, the low complexity NN decoder is able to provide acceptable decoding performance under disguised jamming for high SNRs.

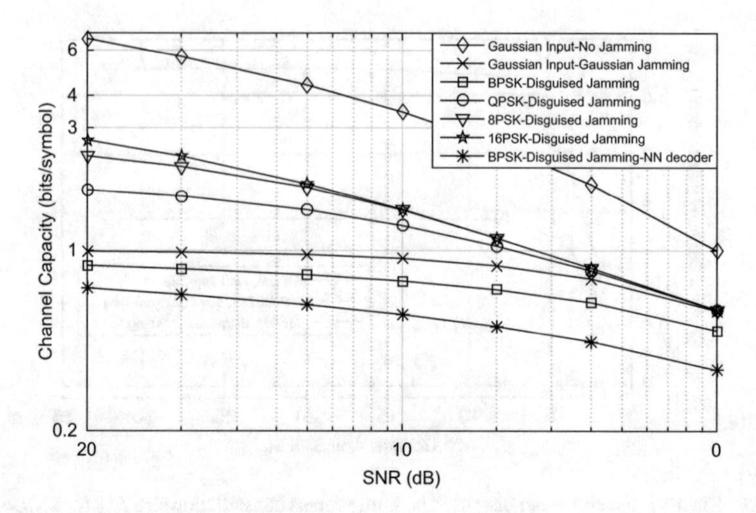

Fig. 5.7 Channel capacity of SP-OFDM versus SNR under different schemes, SJR = 0 dB under disguised jamming

5.5 Discussion on the Worst Jamming Distribution for SP-OFDM

From our previous discussions, we can see that *for an AWGN channel* $R = S + J + N$, *when (i) the constellation of the authorized signal,* Φ, *is fixed; (ii)* S *is uniformly distributed over* Φ; *and (iii) no secure symbol-level precoding is involved, then the worst jamming is the disguised jamming that has an identical distribution with S. In this case, the deterministic code capacity is zero. That is, the traditional OFDM has zero deterministic code capacity under disguised jamming.*

However, in this section, we will show that for systems with secure precoding, i.e.,

$$R = S + e^{j\Theta} J + N,$$

disguised jamming is no longer the worst jamming in terms of channel capacity.

From Lemma 5.1, with the AES-based phase randomization, the channel capacity equals the maximin mutual information between S and R, $I(S, R)$. In this section, we consider the maximin problem (5.19),

$$\begin{aligned} C = & \max_{\mathcal{P}_S} \min_{F_J} I(S, R), \\ s.t. \quad & \int_{\mathbb{C}} |\boldsymbol{x}|^2 \mathrm{d}F_J(\boldsymbol{x}) \le P_J, \end{aligned} \tag{5.46}$$

in practical communication systems, where the transmitted symbol is uniformly distributed in the alphabet and the jamming interference is subject to a finite peak power constraint.

Let the transmitting alphabet $\Phi = \{s_1, s_2, \ldots, s_{M_\Phi}\}$. The size of Φ is $|\Phi| = M_\Phi$. In realistic communication systems, S is uniformly distributed over Φ, i.e.,

$$\Pr\{S = s_i\} = \frac{1}{M_\Phi}, \quad i \in \{1, 2, \ldots, M_\Phi\}. \tag{5.47}$$

Hence, the calculation of the capacity in (5.19) is reduced to finding the CDF F_J of the jamming interference that minimizes the MI $I(S, R)$.

The phase shift Θ is controlled by the authorized user and is uniformly distributed over $\{\frac{2\pi l}{M} \mid l = 0, 1, \ldots, M-1\}$. When M is sufficiently large (e.g., $M \gg \frac{2\pi}{\sigma_\Theta}$, where σ_Θ^2 is the variance of the phase noise existing in practical communication systems), taking the noise effect into consideration, we can approximately model Θ as a continuous RV uniformly distributed over $[0, 2\pi)$ and independent of J. In this way, the phase item within J can be completely absorbed into Θ. As a result, we only need to find the CDF of $|J|$ that minimizes the MI in (5.19). Without loss of generality, in the following, J is degraded to a RV over $\mathbb{R}$.

Let $\mathcal{F}_a$ be the set of all the possible CDFs $F_J(\cdot)$ of $J \in \mathbb{R}$ satisfying the peak power constraint a, that is,

$$\mathcal{F}_a = \{F(\cdot) \mid F(0^-) = 0, F(a) = 1, F(x) \text{ is nondecreasing and right-continuous}\},$$

and it is a convex set. Note that the MI $I(S, R) = H(S) - H(S|R)$ and S is uniformly distributed over Φ. Define functional $G(\cdot)$ as

$$G(F_J) \triangleq -H(S|R). \tag{5.48}$$

Given that S is uniformly distributed over Φ, the maximin problem (5.19) can be reduced to

$$\begin{aligned} &\min_{F_J(\cdot)\in\mathcal{F}_a} \; G(F_J) \\ &\quad s.t. \; \int_0^\infty x^2 \mathrm{d}F_J(x) \le P_J \end{aligned} \tag{5.49}$$

The optimal solution to (5.49) is the worst jamming distribution that results in the minimal capacity from the information theoretic point of view. In the following, we will discuss the worst jamming distribution F_J to (5.49) following the approaches in [31–33].

5.5.1 *Existence of the Worst Jamming Distribution*

In this subsection, we will prove the existence of the worst jamming distribution subject to the power constraints. First we derive the expression of $G(F_J)$ given $F_J \in \mathcal{F}_a$. Define function $u(x, y)$ as

$$u(x, y) \triangleq \frac{1}{\pi\sigma^2} e^{-\frac{x^2+y^2}{\sigma^2}} I_0\left(\frac{2xy}{\sigma^2}\right), \; x \ge 0, y \ge 0, \tag{5.50}$$

where $I_0(\cdot)$ is the modified Bessel function of the first kind with order 0 [34]. Then, for a given noise power σ^2, the conditional PDF of R given S and J can be calculated as

$$f_{R|S,J}(\boldsymbol{r} \mid \boldsymbol{s}_i, x) = u(|\boldsymbol{r} - \boldsymbol{s}_i|, x), \boldsymbol{r} \in \mathbb{C}, \boldsymbol{s}_i \in \Phi, x \geq 0. \tag{5.51}$$

The PDF of R and the posterior distribution of S are

$$f_R(\boldsymbol{r}) = \frac{\sum_i f_{R|S}(\boldsymbol{r} \mid \boldsymbol{s}_i)}{M_\Phi} = \frac{\sum_i \int_0^\infty u(|\boldsymbol{r} - \boldsymbol{s}_i|, x)\mathrm{d}F_J(x)}{M_\Phi}, \tag{5.52}$$

$$\Pr\{S = \boldsymbol{s}_i \mid R = \boldsymbol{r}\} = \frac{f_{R|S}(\boldsymbol{r} \mid \boldsymbol{s}_i)\Pr\{S = \boldsymbol{s}_i\}}{f_R(\boldsymbol{r})} = \frac{\int_0^\infty u(|\boldsymbol{r} - \boldsymbol{s}_i|, x)\mathrm{d}F_J(x)}{M_\Phi \cdot f_R(\boldsymbol{r})}, \tag{5.53}$$

respectively. Define functionals $Q_i(\boldsymbol{r}, F)$ and $L_i(\boldsymbol{r}, F)$ over $\mathbb{C} \times \mathcal{F}_a$ as

$$Q_i(\boldsymbol{r}, F) = \int_0^\infty u(|\boldsymbol{r} - \boldsymbol{s}_i|, x)\mathrm{d}F(x), \ \boldsymbol{r} \in \mathbb{C}, F \in \mathcal{F}_a. \tag{5.54}$$

$$L_i(\boldsymbol{r}, F) = \log\left(\frac{Q_i(\boldsymbol{r}, F)}{\sum_k Q_k(\boldsymbol{r}, F)}\right), \ \boldsymbol{r} \in \mathbb{C}, F \in \mathcal{F}_a. \tag{5.55}$$

It follows from (5.48) that $G(F_J)$ can be expressed as

$$G(F_J) = \frac{1}{M_\Phi}\sum_i \int_{\mathbb{C}} Q_i(\boldsymbol{r}, F_J)L_i(\boldsymbol{r}, F_J)\mathrm{d}\boldsymbol{r}. \tag{5.56}$$

Moreover, $I(S, R)$ is a convex function w.r.t. $F_J(\cdot)$, so is $G(\cdot)$, that is, $G((1-\lambda)F_1 + \lambda F_2) \leq (1-\lambda)G(F_1) + \lambda G(F_2)$ for any $\lambda \in [0, 1]$ and $F_1, F_2 \in \mathcal{F}_a$.

For any $P \geq 0$, define functional $K_P(\cdot)$ as

$$K_P(F) \triangleq \int_0^\infty x^2\mathrm{d}F(x) - P. \tag{5.57}$$

Let $P = P_J$ be the average jamming power constraint, and define

$$\Omega \triangleq \{F(\cdot) \mid F(\cdot) \in \mathcal{F}_a, K_{P_J}(F) \leq 0\}. \tag{5.58}$$

The optimization problem (5.49) can be expressed equivalently as

$$\min_{F \in \Omega} G(F). \tag{5.59}$$

In the following, the subscript J in the distribution function and the average power constraint is discarded for brevity. Next, we will show that functional $G(F)$ can achieve its minimum on Ω. More specifically, we have:

Theorem 5.4 *Given* $\Omega = \{F(\cdot) \mid F(\cdot) \in \mathcal{F}_a, K_P(F) \leq 0\}$, *where* P *is the average jamming power constraint and* $G(F) = -H(S \mid R)$ *on set* Ω, *the real-valued functional* $G(\cdot)$ *can achieve its minimum on* Ω.

The proof of Theorem 5.4 is based on the following Lemma, which makes use of the weak* topology [16] on the set of CDFs over $\mathbb{R}$.

Lemma 5.3 ([32]) *If* G *is a real-valued, weak* continuous functional*[1] *on a weak* compact set*[2] Ω, *then* G *achieves its minimum on* Ω.

According to Lemma 5.3, to prove Theorem 5.4, it is sufficient to prove that Ω is weak* compact and $G(\cdot)$ is weak* continuous. The weak* compactness of Ω was proved in [32]. The weak* continuity of $G(\cdot)$ can be proved by the Lebesgue dominated convergence theorem [35].

Theorem 5.4 establishes the existence of the worst jamming distribution over Ω. Next, we will further show that the worst jamming distribution is also unique.

Corollary 5.1 *The jamming distribution* $F^* \in \Omega$ *that minimizes functional* $G(\cdot)$ *is unique.*

To prove this, assume two jamming distributions $F_0^*, F_1^* \in \Omega$ both can minimize $G(\cdot)$; it can be proved that F_0^*, F_1^* render the same conditional distribution of R given S following [36, Theorem 2.6.3]. Then, by exploiting the characteristic functions of R and J, it follows that F_0^* should be equivalent to F_1^*, which proves the uniqueness of the worst jamming distribution.

To solve the constrained optimization problem, we use the Kuhn-Tucker theorem [32, 37].

Kuhn-Tucker Theorem Let G and K be two convex functionals defined on a convex domain $\mathcal{F}$, $G, K : \mathcal{F} \to \mathbb{R}$. Assume there exists a point $F \in \mathcal{F}$ such that $K(F) < 0$. Let

$$C = \inf_{F \in \mathcal{F}, K(F) \leq 0} G(F), \tag{5.60}$$

and assume $C > -\infty$. Then there exists a constant $\gamma \geq 0$ such that

$$C = \inf_{F \in \mathcal{F}} G(F) + \gamma K(F). \tag{5.61}$$

[1] A functional f defined on X^* is weak* continuous iff for each $x^* \in X^*$ and each neighborhood V of $f(x^*)$, there is a neighborhood U of x^* such that $f(U) \subset V$. Here, X^* denotes the set of CDFs over $\mathbb{R}$.

[2] A set $K \subset X^*$ is said to be weak* compact iff every infinite sequence from K contains a weak* convergent subsequence; an infinite sequence $\{x_n^* \mid n = 0, 1, \cdots\} \subset X^*$ weak* converges to x^* iff $\lim_{n \to \infty} x_n^*(x) = x^*(x), \forall x \in \mathbb{R}$.

Furthermore, if the infimum of (5.60) is achieved at $F_0 \in \mathcal{F}$, then the infimum of (5.61) is also achieved at F_0 and $\gamma K(F_0) = 0$.

Then the following theorem can be obtained.

Theorem 5.5 *There exists a unique solution to the constrained convex optimization problem* (5.49) *on* Ω. *If* $F_0(\cdot) \in \Omega$ *is the solution to* (5.49), *then there exists a* $\gamma \geq 0$ *such that* $F_0(\cdot)$ *is also the solution to the Lagrangian dual problem*

$$\min_{F \in \mathcal{F}_a} G(F) + \gamma K_P(F) \tag{5.62}$$

and $\gamma K_P(F_0) = 0$.

Next we analyze the necessary and sufficient conditions for the minima in the dual problem. Using the definition of weak differentiability [37] in functional analysis, the following result can be obtained.

Theorem 5.6 *Define functional* $g(\cdot\,;\ \cdot)$ *over* $[0, \infty) \times \mathcal{F}_a$ *as*

$$g(x;\, F) \triangleq \frac{\sum_i \int_{\mathbb{C}} u(|\mathbf{r} - \mathbf{s}_i|, x) L_i(\mathbf{r}, F) d\mathbf{r}}{M_\Phi}, \quad x \geq 0,\ F \in \mathcal{F}_a. \tag{5.63}$$

A necessary and sufficient condition for $F_0(\cdot) \in \mathcal{F}_a$ *to be the minima of the dual problem* (5.62) *is,*

$$\int_0^\infty [g(x;\, F_0) + \gamma x^2] dF(x) \geq G(F_0) + \gamma \int_0^\infty x^2 dF_0(x), \ \ \forall F(\cdot) \in \mathcal{F}_a. \tag{5.64}$$

Moreover, if γ *satisfies* $\gamma K_P(F_0) = 0$, *then* F_0 *is also the minima of the primal problem* (5.49).

Furthermore, the following conditions on F_0 can be derived from (5.64) following the approach in [32].

Corollary 5.2 *Let* $E_0 \subseteq [0, a]$ *be the set of points of increase*[3] *of a distribution function* $F_0(\cdot) \in \mathcal{F}_a$.

$$\int_0^\infty [g(x;\, F_0) + \gamma x^2] dF(x) \geq G(F_0) + \gamma \int_0^\infty x^2 dF_0(x), \ \ \forall F(\cdot) \in \mathcal{F}_a, \tag{5.65}$$

iff

$$g(x;\, F_0) + \gamma x^2 \geq G(F_0) + \gamma \int_0^\infty t^2 dF_0(t), \ \ \forall x \in [0, a], \tag{5.66}$$

[3] A point $x_0 \in [0, a]$ is said to be a point of increase of the CDF function $F_0(\cdot)$ if $\exists \delta > 0$, such that $\int_{x_0-\varepsilon}^{x_0+\varepsilon} \mathrm{d}F_0(x) > 0,\ \forall 0 < \varepsilon \leq \delta$.

and

$$g(x; F_0) + \gamma x^2 = G(F_0) + \gamma \int_0^{\infty} t^2 dF_0(t), \;\; \forall x \in E_0. \tag{5.67}$$

In the following, we refer to (5.66) and (5.67) as the **KT conditions**.

5.5.2 Discreteness of the Worst Jamming Distribution

In this subsection, we show that to satisfy (5.66) and (5.67), the worst jamming distribution should be discrete with a finite number of points of increase. To facilitate the proof, we first derive the following asymptotic lower bound of $g(x; F)$ on $x \in \mathbb{R}^+$ for any valid $F(\cdot) \in \mathcal{F}_a$.

Proposition 5.1 *For any $F(\cdot) \in \mathcal{F}_a$ and $x \in \mathbb{R}^+$ sufficiently large, $g(x; F)$ satisfies*

$$g(x; F) \geq -\mathcal{O}(x). \tag{5.68}$$

Proof For any valid $\boldsymbol{r} \in \mathbb{C}$ and $F(\cdot) \in \mathcal{F}_a$, it can be derived that

$$\frac{Q_k(\boldsymbol{r} + \boldsymbol{s}_i, F)}{Q_i(\boldsymbol{r} + \boldsymbol{s}_i, F)} \leq \exp\left(C_0 + C_1|\boldsymbol{r}|\right), \tag{5.69}$$

for some constant C_0, C_1. With (5.69), a lower bound of $L_i(\boldsymbol{r} + \boldsymbol{s}_i, F)$ is

$$L_i(\boldsymbol{r} + \boldsymbol{s}_i, F) = -\log \frac{\sum_k Q_k(\boldsymbol{r} + \boldsymbol{s}_i, F)}{Q_i(\boldsymbol{r} + \boldsymbol{s}_i, F)} \geq C_2 + C_3|\boldsymbol{r}|, \tag{5.70}$$

for some constant C_2, C_3. From (5.50), (5.63) and (5.70), a lower bound of $g(x; F)$ can be found as

$$\begin{aligned} g(x; F) &= \frac{1}{M_\Phi} \sum_i \int_{\mathbb{C}} u(|\boldsymbol{r}|, x) L_i(\boldsymbol{r} + \boldsymbol{s}_i, F) \mathrm{d}\boldsymbol{r} \\ &\geq \frac{1}{\pi\sigma^2} \int_{\mathbb{C}} e^{-\frac{|\boldsymbol{r}|^2 + x^2}{\sigma^2}} I_0\left(\frac{2x|\boldsymbol{r}|}{\sigma^2}\right)(C_2 + C_3|\boldsymbol{r}|) \mathrm{d}\boldsymbol{r}. \end{aligned} \tag{5.71}$$

Moreover,

$$\int_{\mathbb{C}} e^{-\frac{|\boldsymbol{r}|^2 + x^2}{\sigma^2}} I_0\left(\frac{2x|\boldsymbol{r}|}{\sigma^2}\right) \mathrm{d}\boldsymbol{r} = \pi e^{-\frac{x^2}{\sigma^2}} \int_0^{+\infty} e^{-\frac{\rho}{\sigma^2}} I_0\left(\frac{2x\sqrt{\rho}}{\sigma^2}\right) \mathrm{d}\rho, \tag{5.72}$$

$$\int_{\mathbb{C}} e^{-\frac{|\boldsymbol{r}|^2 + x^2}{\sigma^2}} I_0\left(\frac{2x|\boldsymbol{r}|}{\sigma^2}\right) |\boldsymbol{r}| \mathrm{d}\boldsymbol{r} = \pi e^{-\frac{x^2}{\sigma^2}} \int_0^{+\infty} \sqrt{\rho} e^{-\frac{\rho}{\sigma^2}} I_0\left(\frac{2x\sqrt{\rho}}{\sigma^2}\right) \mathrm{d}\rho. \tag{5.73}$$

By exploiting the Whittaker M function [38, 6.643], when x is sufficiently large, (5.72) and (5.73) scale as

$$\begin{aligned} e^{-\frac{x^2}{\sigma^2}} \int_0^{+\infty} e^{-\frac{\rho}{\sigma^2}} I_0(\tfrac{2x\sqrt{\rho}}{\sigma^2}) \mathrm{d}\rho &= \mathcal{O}(1), \\ e^{-\frac{x^2}{\sigma^2}} \int_0^{+\infty} \sqrt{\rho} e^{-\frac{\rho}{\sigma^2}} I_0(\tfrac{2x\sqrt{\rho}}{\sigma^2}) \mathrm{d}\rho &= \mathcal{O}(x). \end{aligned} \tag{5.74}$$

Following (5.71), (5.72), (5.73), and (5.74), we can see that when $x \in \mathbb{R}^+$ is sufficiently large, $g(x; F) \geq -\mathcal{O}(x)$, which completes the proof. ■

With Proposition 5.1, it can be proved that the worst jamming distribution should be discrete and has a finite number of points of increase.

Theorem 5.7 *Let E^* denote the set of points of increase for the optimal jamming distribution F^*, then E^* has a finite number of elements.*

Proof We prove the theorem by contradiction. First we show that if $|E^*| = \infty$, then for the worst jamming distribution F^*,

$$g(x; F^*) + \gamma x^2 \equiv G(F^*) + \gamma \int_0^{\infty} t^2 \mathrm{d}F^*(t), \quad \forall x \in \mathbb{R}^+. \tag{5.75}$$

In fact, since $E^* \subseteq [0, a]$, using the Bolzano-Weierstrass theorem, we can find an infinite sequence x_n, $n = 1, 2, \ldots, \infty$, in E^* such that

$$\lim_{n\to\infty} x_n = x^*, \text{ and } x^* \in [0, a]. \tag{5.76}$$

From Corollary 5.2, we have $g(x_n; F^*) + \gamma x_n^2 \equiv G(F^*) + \gamma \int_0^{\infty} t^2 \mathrm{d}F^*(t), \forall x_n$.

It can be verified that the function defined in (5.66) and (5.67),

$$g(z; F^*) + \gamma z^2, \tag{5.77}$$

is analytic w.r.t. $z \in \mathbb{C}$ for any valid CDF function $F(\cdot) \in \mathcal{F}_a$, which implies its continuity on $[0, a]$. So we have

$$\lim_{n\to\infty} [g(x_n; F^*) + \gamma x_n^2] = g(x^*; F^*) + \gamma (x^*)^2 = G(F^*) + \gamma \int_0^{\infty} t^2 \mathrm{d}F^*(t). \tag{5.78}$$

From the identity theorem [39], if two analytic functions are identical on an infinite set of points in a region along with their limit point, these two functions are identical over the entire region. Therefore,

$$g(x; F^*) + \gamma x^2 - G(F^*) - \gamma \int_0^{\infty} t^2 \mathrm{d}F^*(t) \equiv 0, \forall x \in \mathbb{R}^+. \tag{5.79}$$

which equals (5.75).

Note that $\gamma \geq 0$; next, we consider two cases: (1) $\gamma > 0$ and (2) $\gamma = 0$.

For $\gamma > 0$, note that $g(x; F^*) \geq -\mathcal{O}(x)$; we have

$$g(x; F^*) + \gamma x^2 - G(F^*) - \gamma \int_0^\infty t^2 \mathrm{d}F^*(t) \to \infty \quad \text{as} \quad x \to \infty, \tag{5.80}$$

which contradicts with (5.79). So for $\gamma > 0$, (5.75) cannot hold.

For $\gamma = 0$, the average power constraint is inactive. If for some finite peak power constraint a_0, the worst jamming distribution $F_{a_0}{}^*$ has an infinite number of points of increase, it follows from (5.79) that

$$g(x; F_{a_0}{}^*) - G(F_{a_0}{}^*) \equiv 0, \forall x \in \mathbb{R}^+ \tag{5.81}$$

On the other hand, for any $\alpha > a_0$, $F_{a_0}{}^*(\cdot) \in \mathcal{F}_\alpha$, and the KT conditions are always satisfied by $F_{a_0}{}^*$. Thus, for any $\alpha \geq a_0$, $F_{a_0}{}^*$ can minimize $G(\cdot)$ within $\mathcal{F}_\alpha$, and

$$\lim_{\alpha \to \infty} G(F_\alpha{}^*) = G(F_{a_0}{}^*), \tag{5.82}$$

where $F_\alpha{}^*$ is the worst jamming distribution over $\mathcal{F}_\alpha$. According to Theorem 5.3, when there is no peak jamming power constraint, the MI $I(S, R)$ can be arbitrarily small as the average jamming power P_J increases. Then, from the continuity of $G(\cdot)$, as the peak power constraint α increases, the MI $I(S, R)$ should also be arbitrarily small under the worst jamming distribution. Therefore

$$\lim_{\alpha \to \infty} G(F_\alpha{}^*) = -\log M_\Phi. \tag{5.83}$$

This indicates that $I(S, R) = 0$ for jamming distribution $F_{a_0}{}^*$, which is impossible because the channel capacity of a non-symmetrizable AVC channel is positive as shown in Theorem 5.3. Therefore, the distribution function $F_{a_0}{}^*$ does not exist.

As (5.75) cannot hold for any $\gamma \geq 0$, it implies that E^* only has a finite number of elements, which completes the proof. ■

In the following subsection, we will calculate the worst jamming distribution under different system parameters and demonstrate the result in Theorem 5.7 through numerical examples.

5.5.3 *Numerical Results*

The KT conditions in Corollary 5.2 provide the gradients in the optimization problem [32], and the convexity of functional $G(\cdot)$ guarantees the convergence to the global optimum in the optimization. In this section, we compute the worst jamming distribution using the steepest descent algorithm [32] with different system parameters. We will see that *the worst jamming distribution is discrete in amplitude*

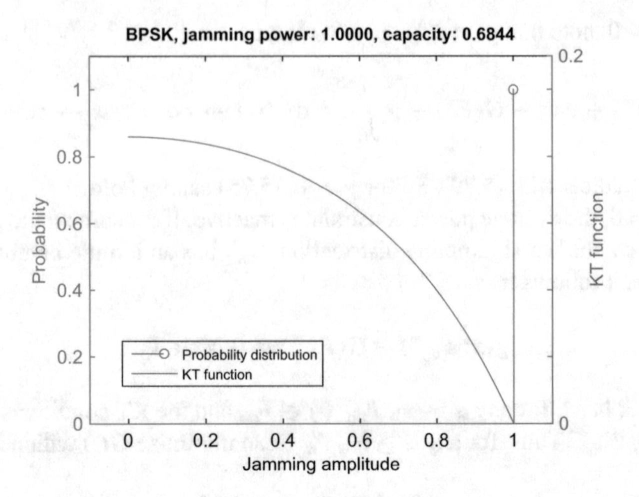

Fig. 5.8 The optimal jamming distribution and KT function versus the jamming amplitude for $a = 1, \gamma = 0$, SNR = 6 dB, BPSK alphabet. The average jamming power is 1 and the corresponding channel capacity is 0.6844

with a finite number of mass points. More specifically, in the following numerical examples, we vary the parameters to study their effects on the worst jamming distribution. Throughout the numerical results, the average power of the legitimate signal is normalized to 1. In each example, we show the worst jamming distribution and the corresponding KT function[4] to verify its optimality.

Example 1 (The worst jamming distribution when the jamming power equals the signal power) In this example, the jammer has the same power constraints as the legitimate transmitter. We set the constellation of the legitimate signal, Φ, to be M_Φ-ary PSK alphabets, where M_Φ is varied in simulation. Hence, the peak power constraint of the jammer is $a = 1$ and the average power constraint is inactive. The SNR is set to be 6 dB. For BPSK and QPSK alphabets, the worst jamming distributions and KT functions are plotted in Figs. 5.8 and 5.9, respectively. For the BPSK case, the worst jamming distribution is quite special, where the jammer uses a single tone jamming with the maximal peak power. In the QPSK case, the worst jamming distribution has two mass points in amplitude. The underlying argument is that, to minimize the mutual information between S and R, jamming with the maximal available power is not always the best practice.

Example 2 (The effect of peak and average power constraints) In this example, we show the effect of peak and average power constraints on the worst jamming distribution. BPSK is employed by the legitimate transmitter and the SNR is 6 dB.

[4]The KT function for a jamming distribution F is $KT(x) = g(x; F) + \gamma x^2 - G(F) - \gamma \int_0^\infty t^2 \mathrm{d}F(t)$, where x is the jamming amplitude.

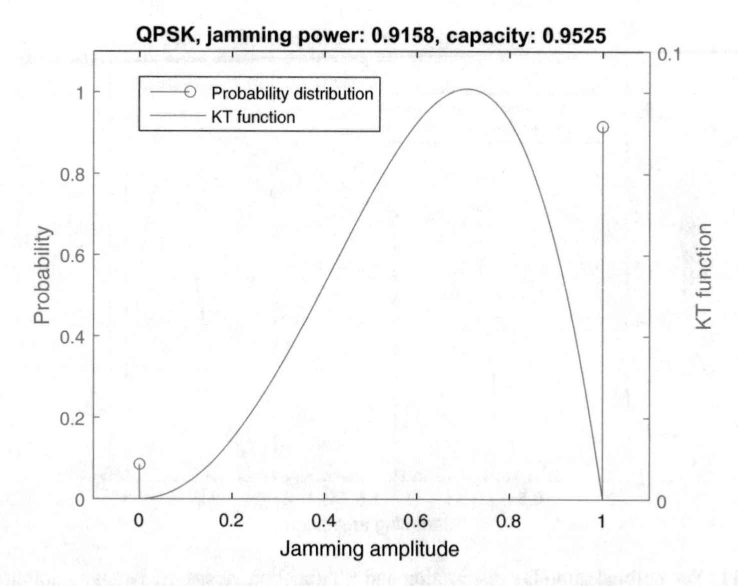

Fig. 5.9 The optimal jamming distribution and KT function versus the jamming amplitude for $a = 1, \gamma = 0$, SNR = 6 dB, QPSK alphabet. The average jamming power is 0.9158 and the corresponding channel capacity is 0.9525

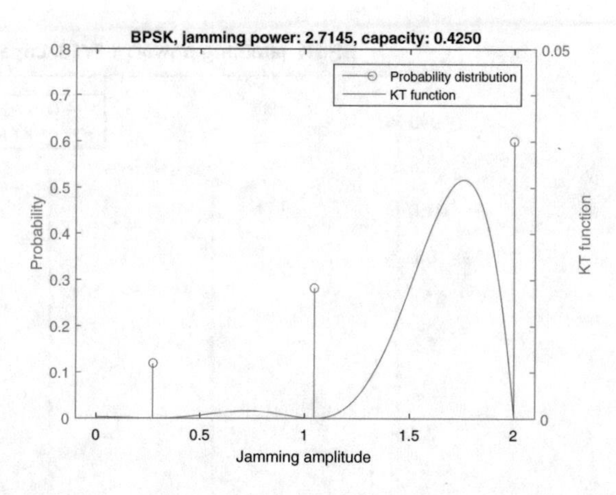

Fig. 5.10 The optimal jamming distribution and KT function versus the jamming amplitude for $a = 2, \gamma = 0$, SNR = 6 dB, BPSK alphabet. The average jamming power is 2.7145 and the corresponding channel capacity is 0.425

We first set the average power constraint of the jammer to be inactive and vary the peak power constraint $a = 1, 2, 3$. The resulting probability distributions and KT functions are shown in Figs. 5.8, 5.10 and 5.11. Then we fix the peak power constraint $a = 2$ and vary the Lagrangian multiplier $\gamma = 0, 0.2, 0.4$ to regulate the average jamming power constraint, whose results are plotted in Figs. 5.10, 5.12 and 5.13. It is shown that the optimal jamming distribution varies a lot with different peak and average power constraints. It can also be observed that, when the average power constraint is inactive, with the increase of the peak jamming power a, the number of mass points in the optimal jamming distribution also increases.

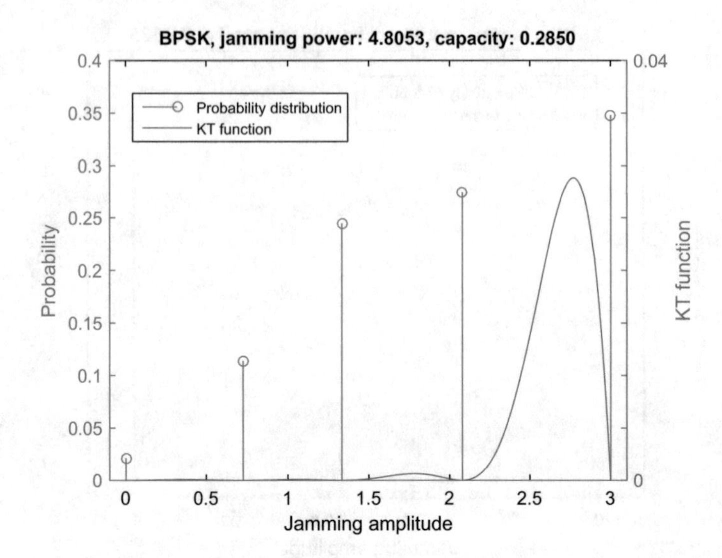

Fig. 5.11 The optimal jamming distribution and KT function versus the jamming amplitude for $a = 3, \gamma = 0$, SNR = 6 dB, BPSK alphabet. The average jamming power is 4.8053 and the corresponding channel capacity is 0.285

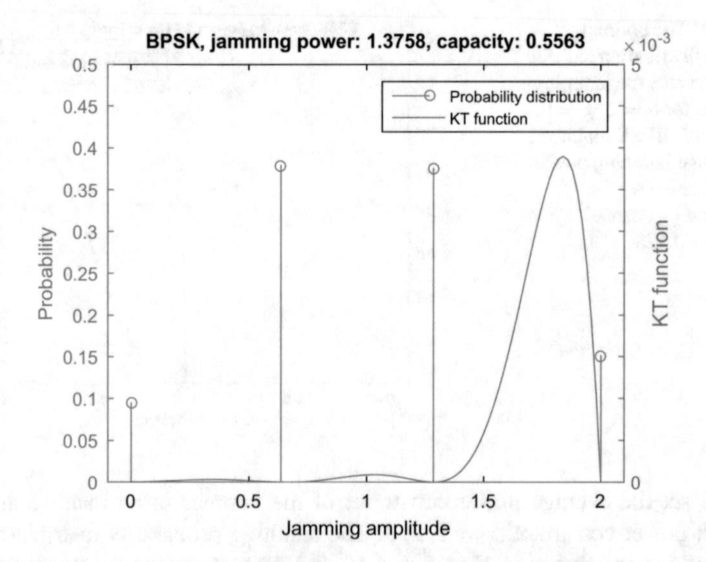

Fig. 5.12 The optimal jamming distribution and KT function versus the jamming amplitude for $a = 2, \gamma = 0.2$, SNR = 6 dB, BPSK alphabet. The average jamming power is 1.3758 and the corresponding channel capacity is 0.5563

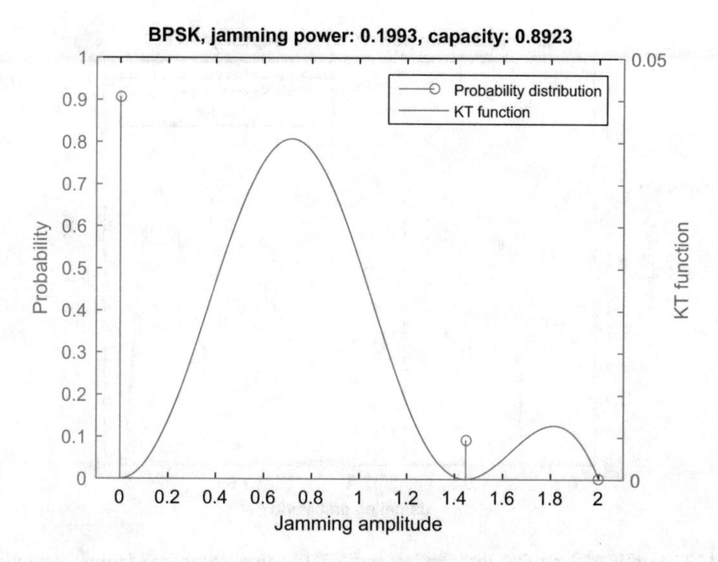

Fig. 5.13 The optimal jamming distribution and KT function versus the jamming amplitude for $a = 2, \gamma = 0.4$, SNR = 6 dB, BPSK alphabet. The average jamming power is 0.1993 and the corresponding channel capacity is 0.8923

Example 3 (The effect of constellation size) In this example, we compute the worst jamming distribution under different constellation size. The M_Φ-ary PSK alphabets are considered, where $M_\Phi = 2, 4$ and 8. We fix the peak power constraint $a = 2$, the Lagrangian multiplier $\gamma = 0.4$ and SNR = 6 dB. The results are shown in Figs. 5.13, 5.14 and 5.15. We can observe that, by changing the alphabet from QPSK to 8PSK, the variation on the worst jamming distribution is minor. This indicates that, for the M_Φ-ary PSK modulation, given the jamming power constraints and SNR level, the worst jamming distribution converges with the increase of constellation size M_Φ.

Example 4 (The effect of SNR level) In this example, we examine the effect of SNR level on the worst jamming distribution. We fix the peak power constraint $a = 2$ and Lagrangian multiplier $\gamma = 0$, i.e., the average jamming power constraint is inactive. BPSK alphabet is employed by the legitimate signals. The SNR level is varied over 10, 6 and 3 dB, where the corresponding results are shown in Figs. 5.10, 5.16, and 5.17, respectively. It is shown that the worst jamming distribution varies a lot with the changes of the SNR level. It can be observed that, with the increase of SNR level, the number of mass points in the worst jamming distribution also increases.

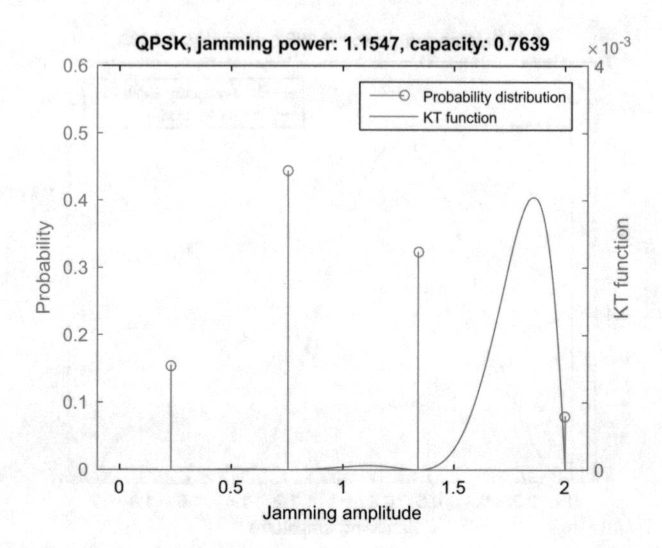

Fig. 5.14 The optimal jamming distribution and KT function versus the jamming amplitude for $a = 2, \gamma = 0.4$, SNR = 6 dB, QPSK alphabet. The average jamming power is 1.1547 and the corresponding channel capacity is 0.7639

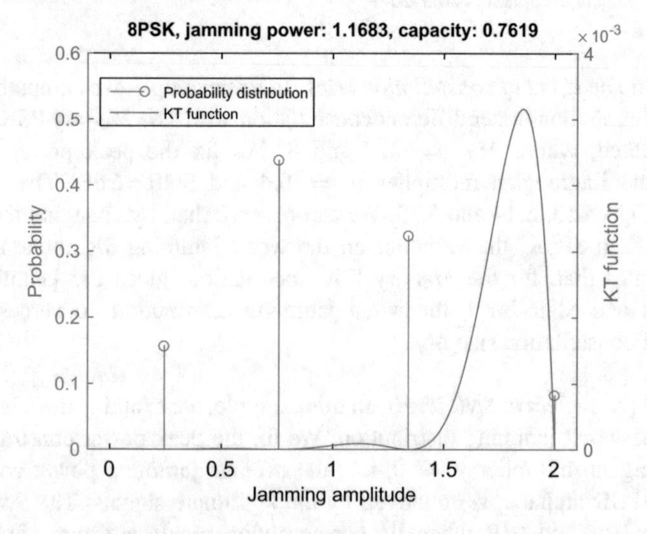

Fig. 5.15 The optimal jamming distribution and KT function versus the jamming amplitude for $a = 2, \gamma = 0.4$, SNR = 6 dB, 8PSK alphabet. The average jamming power is 1.1683 and the corresponding channel capacity is 0.7619

5.6 Conclusions

In this chapter, we presented a highly secure and efficient OFDM system under disguised jamming, named securely precoded OFDM (SP-OFDM), by exploiting secure symbol-level precoding basing on phase randomization. We analyzed the

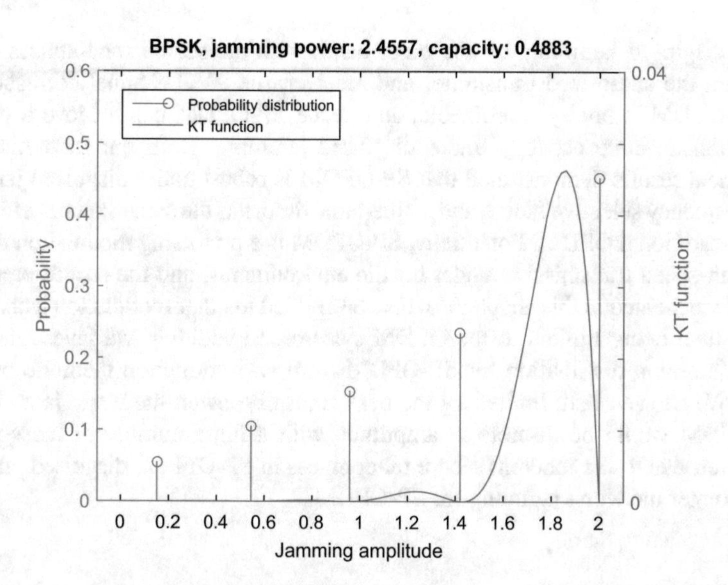

Fig. 5.16 The optimal jamming distribution and KT function versus the jamming amplitude for $a = 2, \gamma = 0$, SNR = 10 dB, BPSK alphabet. The average jamming power is 2.4557 and the corresponding channel capacity is 0.4883

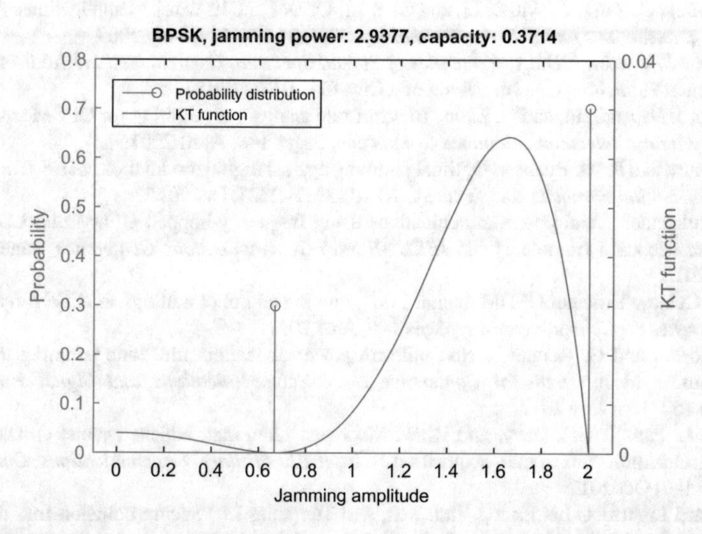

Fig. 5.17 The optimal jamming distribution and KT function versus the jamming amplitude for $a = 2, \gamma = 0$, SNR = 3 dB, BPSK alphabet. The average jamming power is 2.9377 and the corresponding channel capacity is 0.3714

channel capacity of the traditional OFDM and the innovative SP-OFDM under hostile jamming using the arbitrarily varying channel (AVC) model. It was shown that the deterministic code capacity of the traditional OFDM is zero under the worst disguised jamming; on the other hand, with the secure randomness shared between the authorized transmitter and receiver, the AVC channel corresponding to SP-OFDM is not symmetrizable, and hence SP-OFDM can achieve a positive deterministic code capacity under disguised jamming. Both our theoretical and numerical results demonstrated that SP-OFDM is robust under disguised jamming and frequency selective fading and, at the same time, has the same spectral efficiency as the traditional OFDM. Potentially, SP-OFDM is a promising modulation scheme for high-speed transmission under hostile environments, and the secure precoding scheme presented in this chapter can also be applied to other modulation techniques, rather than being limited to the OFDM systems. In addition, we investigated the worst jamming distribution for SP-OFDM from an information theoretic point of view. We showed that, limited by the peak transmit power, the worst jamming for SP-OFDM would be discrete in amplitude with a finite number of mass points. However, due to the inherent secure randomness in SP-OFDM, disguised jamming is no longer the worst jamming for SP-OFDM.

References

1. T. Hwang, C. Yang, G. Wu, S. Li, and G. Y. Li. OFDM and its wireless applications: A survey. *IEEE Transactions on Vehicular Technology*, 58(4):1673–1694, May 2009.
2. Andrew J. Viterbi. *CDMA: Principles of Spread Spectrum Communication*. Addison Wesley Longman Publishing Co., Inc., Redwood City, CA, USA, 1995.
3. L. Jun, J. H. Andrian, and C. Zhou. Bit error rate analysis of jamming for OFDM systems. In *2007 Wireless Telecommunications Symposium*, pages 1–8, April 2007.
4. S. Amuru and R. M. Buehrer. Optimal jamming against digital modulation. *IEEE Transactions on Information Forensics and Security*, 10(10):2212–2224, Oct 2015.
5. L. Mailaender. Anti-jam communications using frequency-hopped OFDM and LDPC with erasure decoding (minotaur). In *IEEE Military Communications Conference*, pages 84–88, Nov 2013.
6. T. C. Clancy. Efficient OFDM denial: Pilot jamming and pilot nulling. In *IEEE International Conference on Communications*, pages 1–5, June 2011.
7. P. Cuccaro and G. Romano. Non uniform power allocation pilot tone jamming in OFDM systems. In *International Conference on Telecommunications and Signal Processing*, pages 152–155, July 2017.
8. M. J. L. Pan, T. C. Clancy, and R. W. McGwier. Jamming attacks against OFDM timing synchronization and signal acquisition. In *IEEE Military Communications Conference*, pages 1–7, Oct 2012.
9. Leonard Lightfoot, Lei Zhang, Jian Ren, and Tongtong Li. Secure collision-free frequency hopping for OFDMA-based wireless networks. *EURASIP J. Adv. Signal Process*, 2009:1:1–1:11, March 2009.
10. T. Song, K. Zhou, and T. Li. CDMA system design and capacity analysis under disguised jamming. *IEEE Transactions on Information Forensics and Security*, 11(11):2487–2498, Nov 2016.
11. L. Zhang and T. Li. Anti-jamming message-driven frequency hopping-part ii: Capacity analysis under disguised jamming. *IEEE Transactions on Wireless Communications*, 12(1):80–88, January 2013.
12. T. Song, Z. Fang, J. Ren, and T. Li. Precoding for OFDM under disguised jamming. In *2014 IEEE Global Communications Conference*, pages 3958–3963, Dec 2014.

13. T. Ericson. The noncooperative binary adder channel. *IEEE Transactions on Information Theory*, 32(3):365–374, May 1986.
14. L. Zhang, H. Wang, and T. Li. Anti-jamming message-driven frequency hopping-part i: System design. *IEEE Transactions on Wireless Communications*, 12(1):70–79, January 2013.
15. A. Furuskar, S. Mazur, F. Muller, and H. Olofsson. EDGE: enhanced data rates for GSM and TDMA/136 evolution. *IEEE Personal Communications*, 6(3):56–66, Jun 1999.
16. David G. Luenberger. *Optimization by Vector Space Methods*. John Wiley & Sons, Inc., New York, NY, USA, 1st edition, 1997.
17. Rudolf Ahlswede. Elimination of correlation in random codes for arbitrarily varying channels. *Zeitschrift für Wahrscheinlichkeitstheorie und Verwandte Gebiete*, 44(2):159–175, 1978.
18. David Blackwell, Leo Breiman, and A. J. Thomasian. The capacities of certain channel classes under random coding. *Ann. Math. Statist.*, 31(3):558–567, 09 1960.
19. Frederic P. Miller, Agnes F. Vandome, and John McBrewster. *Advanced Encryption Standard*. Alpha Press, 2009.
20. John R. Barry, David G. Messerschmitt, and Edward A. Lee. *Digital Communication: Third Edition*. Kluwer Academic Publishers, Norwell, MA, USA, 2003.
21. Achim Klenke. *Probability Theory: A Comprehensive Course*. Springer, 2008.
22. I. Csiszar and P. Narayan. The capacity of the arbitrarily varying channel revisited: positivity, constraints. *IEEE Transactions on Information Theory*, 34(2):181–193, Mar 1988.
23. I. Csiszar. Arbitrarily varying channels with general alphabets and states. *IEEE Transactions on Information Theory*, 38(6):1725–1742, Nov 1992.
24. J. Martin Borden, David M. Mason, and Robert J. McEliece. Some information theoretic saddlepoints. *SIAM Journal on Control and Optimization*, 23(1):129–143, 1985.
25. D.Z. Du and P.M. Pardalos. *Minimax and Applications*. Springer US, 1995.
26. Abbas El Gamal and Young-Han Kim. *Network Information Theory*. Cambridge University Press, New York, NY, USA, 2012.
27. A. Morello and V. Mignone. DVB-S2: The second generation standard for satellite broad-band services. *Proceedings of the IEEE*, 94(1):210–227, Jan 2006.
28. Sae-Young Chung, T. J. Richardson, and R. L. Urbanke. Analysis of sum-product decoding of low-density parity-check codes using a Gaussian approximation. *IEEE Transactions on Information Theory*, 47(2):657–670, Feb 2001.
29. Theodore Rappaport. *Wireless Communications: Principles and Practice*. Prentice Hall PTR, Upper Saddle River, NJ, USA, 2nd edition, 2001.
30. A. R. Runnalls. Kullback-leibler approach to Gaussian mixture reduction. *IEEE Transactions on Aerospace and Electronic Systems*, 43(3):989–999, July 2007.
31. S. Shamai and I. Bar-David. The capacity of average and peak-power-limited quadrature Gaussian channels. *IEEE Transactions on Information Theory*, 41(4):1060–1071, Jul 1995.
32. I. C. Abou-Faycal, M. D. Trott, and S. Shamai. The capacity of discrete-time memoryless Rayleigh-fading channels. *IEEE Transactions on Information Theory*, 47(4):1290–1301, May 2001.
33. T. H. Chan, S. Hranilovic, and F. R. Kschischang. Capacity-achieving probability measure for conditionally Gaussian channels with bounded inputs. *IEEE Transactions on Information Theory*, 51(6):2073–2088, June 2005.
34. *NIST Digital Library of Mathematical Functions*. http://dlmf.nist.gov/, Release 1.0.15 of 2017-06-01. F. W. J. Olver, A. B. Olde Daalhuis, D. W. Lozier, B. I. Schneider, R. F. Boisvert, C. W. Clark, B. R. Miller and B. V. Saunders, eds.
35. M. C. Gursoy, H. V. Poor, and S. Verdu. The capacity of the noncoherent Rician fading channel. Technical report, Princeton University, December 2002.
36. Thomas M. Cover and Joy A. Thomas. *Elements of Information Theory*. Wiley-Interscience, New York, NY, USA, 1991.

37. Joel G. Smith. The information capacity of amplitude- and variance-constrained sclar Gaussian channels. *Information and Control*, 18(3):203–219, 1971.
38. I. S. Gradshteyn and I. M. Ryzhik. *Table of integrals, series, and products*. Elsevier/Academic Press, Amsterdam, seventh edition, 2007.
39. D. Sarason. *Complex Function Theory*. American Mathematical Society.

Chapter 6
Multiband Transmission Under Jamming: A Game Theoretic Perspective

6.1 Game Theory and Communication Under Jamming

In this section, we will first introduce the basic principle of game theory (particularly two-party zero-sum games) and then briefly review how it is applied in different areas of communications, especially on jamming resistant multiband communications.

6.1.1 The Concept of Game Theory

Game theory is "the study of mathematical models of conflict and cooperation between intelligent rational decision-makers" [1], which applies to a wide range of behavioral relations, and is now an umbrella term for the science of logical decision-making in various areas, including engineering, economics, and psychology. Modern game theory began with the introduction and investigation on the theory of two-party zero-sum games by John von Neumann, and today it covers a large number of games categorized by different criteria, including but not limited to the number of players, the information available to players, and so on [2]. In this subsection, we will introduce the classical two-party zero-sum game and use it as a tool to investigate the game between jamming and anti-jamming techniques in communications.

Shortly speaking, in a zero-sum game, one person's gains result in losses for the other participants. When the number of participants is limited to only two, it becomes even more straightforward. For a two-party zero-sum game, the payoff function of Player II is the negative of the payoff of Player I, so we only need to consider the single payoff function of Player I.

T. Li et al., *Wireless Communications Under Hostile Jamming: Security and Efficiency*, https://doi.org/10.1007/978-981-13-0821-5_6

Definition 6.1 The strategic form of a two-party zero-sum game is given by a triplet $(\mathcal{X}, \mathcal{Y}, A)$, where

1. $\mathcal{X}$ is a nonempty set, the set of strategies of Player I;
2. $\mathcal{Y}$ is a nonempty set, the set of strategies of Player II;
3. A is a real-valued payoff function defined on $\mathcal{X} \times \mathcal{Y}$.

The interpretation is as follows. As the first step, Player I chooses $x \in X$ and Player II chooses $y \in Y$, each unaware of the choice of the other. Then their choices are made known and Player I wins the amount $A(x, y)$ from Player II. If A is negative, Player I pays the absolute value of this amount to Player II. Thus, $A(x, y)$ represents the winnings of Player I and the losses of Player II.

It should be noted that the strategies could be both *pure* or *mixed*. We refer to elements of X or Y as *pure* strategies. A *mixed* strategy is a combination of pure strategies, which are chosen at random in various proportions. It is also commonly assumed that when a player uses a mixed strategy, he is only interested in his average return. He does not care about his maximum possible winnings or losses-only the average.

Definition 6.2 A saddle point of a two-party zero-sum game is defined as a strategy pair (x^*, y^*) that satisfies

$$A(x^*, y^*) = \max_{x \in \mathcal{X}} \min_{y \in \mathcal{Y}} A(x, y) = \min_{y \in \mathcal{Y}} \max_{x \in \mathcal{X}} A(x, y), \tag{6.1}$$

i.e.,

$$A(x, y^*) \leq A(x^*, y^*) \leq A(x^*, y), \quad \forall x \in \mathcal{X}, y \in \mathcal{Y}, \tag{6.2}$$

where

1. $A(x^*, y^*)$ is the value of the game;
2. x^* is the optimal strategy for Player I such that Player I's gain is at least $A(x^*, y^*)$ no matter what Player II does, and;
3. y^* is the optimal strategy for Player II such that Player II's loss is at most $A(x^*, y^*)$ no matter what Player I does.

Hence, solving a two-party zero-sum game essentially becomes finding the saddle-point strategy pair and the corresponding value of the game as defined above.

6.1.2 Game Theory and Its Applications in Communications

In traditional research on jamming strategy and jamming mitigation, there is generally an assumption that the jammer or the authorized user can access at least part of the information about the transmission pattern of its adversary. As such, the jammer can launch more effective jamming by exploiting the information it has about the authorized user, e.g., correlated jamming [3–5] or disguised jamming [6–10]. For jamming mitigation, the authorized user can mitigate the jammer's effect by applying a particular anti-jamming scheme that is robust against a specific jamming pattern [11, 12]. The underlying assumption is that the jamming varies slowly such that the authorized user has sufficient time to track and react to the jamming.

However, if the jammer is intelligent and can switch its patterns fast enough, then it would be impossible for the authorized user to track and react in real time. In this case, when choosing the strategy to maximize its capacity, the authorized user has no knowledge of the jamming strategy. Similarly, while trying to minimize the capacity of the authorized user, the jammer has no knowledge of the user strategy, either. Regarding this scenario, there has been a surge in research that applies game theory to characterize and analyze strategies for communication systems under jamming with unpredictable strategies.

A lot of work on game theory in communications has been focused on *the single user and single band case* [13–17]. The optimal jamming strategy under the Gaussian test channel was investigated in [13], and the worst additive noise for a communication channel under a covariance constraint was studied in [14]. The capacity of channels with block memory was investigated in [15], which showed that both the optimal coding strategy and the optimal jamming strategy are independent from symbol to symbol within a block. The authors in [16] discussed the minimax game between an authorized user and a jammer for any combinations of "hard" or "soft" input and output quantization with additive noise and average power constraints. In [17], a dynamic game between a communicator and a jammer was considered, where the participants choose their power levels randomly from a finite space subject to temporal energy constraints.

Application of game theory to *multiuser and multiband/multicarrier communications* has been brought to attention in recent years [18–22]. In [18], the authors proposed a decentralized strategy to find out the optimal precoding/multiplexing matrices for a multipoint-to-multipoint communication system composed of a set of wideband links sharing the same physical resources. In [19], a scheme aiming for fair allocation of subcarriers, rates, and power for multiuser orthogonal frequency-division multiple-access (OFDMA) systems was proposed to maximize the overall system rate, subject to each user's maximal power and minimal rate constraints. In [20], jamming mitigation was carried out by maximizing the sum signal-to-interference and noise ratio (SINR) for multichannel communications. In [21], the authors considered a particular scenario where K users and a jammer share a common spectrum of N orthogonal tones and examined how each user could maximize its own total sum rate selfishly. In [22], the authors investigated the secrecy capacity of the users under malicious eavesdropping and friendly jamming.

Game theory has also been applied to *cognitive radios and ad hoc networks* [23–27]. New techniques for analyzing networks of cognitive radios that can alter either their power levels or their signature waveforms through the use of game models were introduced in [23]. In [24], a game theoretic overview of dynamic spectrum sharing was provided regarding analysis of network users' behaviors, efficient dynamic distributed design, and performance optimality. A game theoretic power control framework for spectrum sensing in cognitive radio networks was proposed in [25], and the minimax game for cooperative spectrum sensing in centralized cognitive radio networks was investigated in [26]. In [27], the authors developed a game theoretic framework to construct convergent interference avoidance (IA) algorithms in ad hoc networks with multiple distributed receivers.

6.1.3 Game Theory and Multiband Communications

It has been an active research topic to combat jamming for multiband communications using game theory, since the property of "multiband" provides considerable flexibility to the strategies of both an authorized user and a jammer, who will try the best way to utilize resources (channels and power) such that his own interest is maximized in the game. Here we will review a couple of most relevant research papers in literature.

6.1.3.1 A Bayesian Jamming Game in an OFDM Wireless Network

Garnaev et al. [20] presented a zero-sum game between an authorized transmitter and a jammer. In the game, the transmitter assigns different power levels over n channels in order to maximize his throughput under the worst-case jamming scenario. The interference signal is described as a jammer distributing an extra Gaussian noise of the total power $\bar{J}$ among the channels. The strategy of the transmitter is the vector $\mathbf{T} = (T_1, \ldots, T_n)$ with $T_i \geq 0, i \in [1, n]$, such that $\sum_{i=1}^{n} T_i = \bar{T}$, where $\bar{T} > 0$ is the total available power for the transmitter, and T_i is the power level assigned for channel i. Similarly, the strategy of the jammer is the vector $\mathbf{J} = (J_1, \ldots, J_n)$ with $J_i \geq 0, i \in [1, n]$, such that $\sum_{i=1}^{n} J_i = \bar{J}$, where $\bar{J} > 0$ is the total available power for the jammer. The transmitter's payoff is defined as the signal-to-interference and noise ratio (SINR):

$$v_T(\mathbf{T}, \mathbf{J}) = \sum_{i=1}^{n} \frac{\alpha_i T_i}{\sigma^2 + \beta_i J_i}, \tag{6.3}$$

where α_i (resp. β_i) are the fading channel gains of the transmitter (resp. the jammer) on channel i and σ^2 is the background noise level.

The transmitter's objective is to maximize his payoff, $v_T(\mathbf{T}, \mathbf{J})$, while the jammer's objective is to minimize the transmitter's payoff, i.e., to maximize the function $-v_T(\mathbf{T}, \mathbf{J})$. It is assumed that all the fading channel gains α_i and β_i, the noise level σ^2, and the total powers $\bar{T}$ and $\bar{J}$ are known to both rivals. An important result is given in the literature regarding this game with complete information, which is summarized in Theorem 6.1.

Theorem 6.1 *In the game with complete information about fading channel gains, the equilibrium (saddle point)* $(\boldsymbol{T}^*, \boldsymbol{J}^*) = (T^*(\omega), J^*(\omega))$ *is given as follows:*

$$J_i^*(\omega) = \frac{\alpha_i}{\beta_i}\left[\frac{1}{\omega} - \frac{\sigma^2}{\alpha_i}\right]^+, \quad i \in [1, n], \tag{6.4}$$

and

$$T_i^*(\omega) = \begin{cases} \frac{(\beta_i/\alpha_i)}{\sum_{k \in I^J(\omega)} (\beta_k/\alpha_k)} \bar{T}, & i \in I^J(\omega), \\ 0, & i \notin I^J(\omega), \end{cases} \tag{6.5}$$

where $I^J(\omega) := \{i \in [1, n] : J_i^*(\omega) > 0\}$ *is the set of jammed channels and* ω *is the unique root of the following water-filling equation:*

$$\sum_{i=1}^{n} \frac{\alpha_i}{\beta_i}\left[\frac{1}{\omega} - \frac{\sigma^2}{\alpha_i}\right]^+ = \bar{J}. \tag{6.6}$$

Theorem 6.1 shows that if the payoff function of the two-party zero-sum game is SINR, the optimal strategy for the jammer is a water-filling-like solution as shown in (6.4); for the transmitter, he does not need to allocate power to the unjammed channels but should distribute the available power to the jammed channels according to (6.5). This is a little counterintuitive but can be explained easily from a game-theoretic perspective. That is, the jammer, to limit the transmission of the transmitter, would jam the channels that are favorable to the transmitter, which implies that the condition of unjammed channels is actually not good enough to be utilized by the transmitter in terms of achieving a higher SINR.

6.1.3.2 CSI Usage Over Parallel Fading Channels Under Jamming Attacks: A Game Theory Study

Wei et al. [28] considered a system of three nodes, source (S), destination (D), and jammer (J), where S sends data to D under the jammer's attack. A parallel flat-fading channel model is assumed, where each fading block has M parallel subchannels defined as

$$Y_i[n] = H_i[n]X_i[n] + W_i[n] + Z_i[n], \tag{6.7}$$

for $i = 1, \ldots, M$ and $n = 1, 2, \ldots, LN$, where channel fading coefficients $H_i[n]$ remain constant over $(l-1)N + 1 \le n \le lN$ for any $1 \le l \le L$ and vary independently over different blocks (of N channel uses) and subchannels, satisfying the distribution $H_i[n] \sim \mathcal{CN}(0, \sigma_i^2)$, where $\mathcal{CN}(\mu, \sigma^2)$ denotes the complex circular Gaussian distribution with mean μ and variance σ^2. The M subchannels model the M parallel frequency bands. Each code word spans over $M \times N \times L$ channel uses. Channel additive white Gaussian noise sequences $W_i[n]$ are assumed i.i.d across MNL channel uses with distribution $\mathcal{CN}(0, N_0)$. Jamming signals $Z_i[n]$ are independent across all channel uses. The problem is formulated under long-term average power constraints: $\frac{1}{LN}\sum_{i=1}^{M}\sum_{l=1}^{L}\sum_{n=(l-1)N+1}^{lN}|X_i[n]|^2 \leq P_S$, and $\frac{1}{LN}\sum_{i=1}^{M}\sum_{l=1}^{L}\sum_{n=(l-1)N+1}^{lN}|Z_i[n]|^2 \leq P_J$.

They adopted the average transmission rate as the metric and modeled the interaction between jammer and the legitimate transceiver using a two-party zero-sum game. The strategy space for each player is defined by $C_i = \{\text{Hp}(\{\alpha_j, P_j\}), \text{Sp}(\{P_j\})\}$, for $i \in \{S, J\}$ and $1 \leq j \leq M$, where Hp represents random hopping over M parallel channels, while Sp represents spreading power over an entire frequency band. After each player decides on hopping or spreading, it needs to further decide what is the associated hopping pattern, namely, the hopping probability $\{\alpha_j\}$ and its associated power allocation function $\{P_j\}$ under a given CSI. Therefore, as far as the strategy goes, each player (jammer or transmitter) chooses hopping or spreading, as well as the corresponding hopping patterns or power functions, accordingly.

By finding the optimal strategies for the transmitter and the jammer in the game described above, they answered two important questions: for the transmitter and the jammer, is it a good idea to hop around different subchannels, and how should the power be distributed? The results are summarized in Theorem 6.2.

Theorem 6.2 *The unique Nash equilibrium (saddle point) of the game is* (S_{sp}, J_{sp}) *with transmitter and jammer taking their respective power control functions* $\{P_i^*(\boldsymbol{h})\}$ *and* $\{J_i^*(\boldsymbol{h})\}$ *given below:*

$$P_i^*(\boldsymbol{h}) = \begin{cases} (\frac{1}{\lambda} - \frac{N_0}{h_i})^+, & h_i \leq \Gamma, i = 1, \ldots, M, \\ \frac{h_i}{\lambda(h_i + \lambda/v)}, & h_i > \Gamma, i = 1, \ldots, M, \end{cases} \tag{6.8}$$

and

$$J_i^*(\boldsymbol{h}) = \begin{cases} 0, & h_i \leq \Gamma, i = 1, \ldots, M, \\ \frac{h_i}{v(h_i + \lambda/v)} - N_0, & h_i > \Gamma, i = 1, \ldots, M, \end{cases} \tag{6.9}$$

where $\boldsymbol{h} = \{h_1, \ldots, h_M\}$ *are the channel states,* $\Gamma = \frac{N_0 \lambda}{1 - N_0 v}$ *and* λ *and* v *are two constants determined by the power constraints and are functions of channel statistics.*

Theorem 6.2 reveals that when the channel state information $\{h_1, \ldots, h_M\}$ is available to both the jammer and the transmitter, neither the transmitter nor the jammer will adopt the hopping strategy. When either the jammer or the transmitter deviates from spreading to hopping, it always ends up with a beneficial effect for its opponent. Hence, both of them shall fully exploit all degrees of freedom as much as possible in terms of encoding and spreading their powers optimally as shown in Theorem 6.2.

6.1.3.3 Equilibrium Strategies for an OFDM Network That Might Be Under a Jamming Attack

Garnaev et al. [29] considered a scenario that involves a legitimate transmitter (the user) using n channels to communicate to a receiver. These channels are characterized by n channel gains h_i. An adversary might or might not be present and, if present, has n channel gains g_i to the receiver. The pure strategy for the user is $\mathbf{P} = (P_1, \ldots, P_n)$, where P_i is the power transmitted by the user through channel i, $\sum_{i=1}^{n} P_i = \bar{P}$, and $\bar{P}$ is the total power budget.

The adversary might be present, and the user has a priori knowledge about the adversary's presence, i.e., the probability that the adversary is present is q^1, and the probability that the adversary is not present is $q^0 = 1 - q^1$. If the adversary is present, the pure strategy for the adversary is $\mathbf{J} = (J_1, \ldots, J_n)$, where J_i is the power used by the adversary to jam channel i, $\sum_{i=1}^{n} J_i = \bar{J}$, and $\bar{J}$ is the total jamming power.

The problem is formulated as a classical two-party zero-sum game, where the user wants to maximize his payoff, while the adversary wants to minimize it. The payoff function is calculated as the expected throughput of the user:

$$E(\mathbf{P}, \mathbf{J}) = q^1 \sum_{i=1}^{n} \ln\left(1 + \frac{h_i P_i}{\sigma^2 + g_i J_i}\right) + q^0 \sum_{i=1}^{n} \ln\left(1 + \frac{h_i P_i}{\sigma^2}\right), \tag{6.10}$$

where σ^2 is the non-adversarial background noise power, which is assumed constant across the channels.

They showed that the game formulated above has a unique equilibrium (saddle-point) strategy $(\mathbf{P}^*, \mathbf{J}^*)$ with an arbitrary probability of the adversary's presence. Particularly, in the case that $q^0 = 1$, i.e., the adversary is absent with certainty, then the optimal transmission strategy for the user coincides with the classical water-filling solution maximizing throughput in OFDM transmission, while in the case that $q^1 = 1$, i.e., the adversary is present with certainty, then the equilibrium (saddle-point) strategies are given as follows:

$$P_i^*(w, v) = \begin{cases} \frac{h_i}{h_i v + g_i w} \frac{v}{w}, & i \in I_{11}(w, v), \\ \frac{1}{w} - \frac{\sigma^2}{h_i}, & i \in I_{10}(w, v), \\ 0, & i \in I_{00}(w), \end{cases} \tag{6.11}$$

and

$$J_i^*(w, v) = \begin{cases} \frac{h_i}{h_i v + g_i w} - \frac{\sigma^2}{g_i}, & i \in I_{11}(w, v), \\ 0, & i \in I_{00}(w) \cup I_{10}(w, v), \end{cases} \tag{6.12}$$

where I_{00}, I_{10}, and I_{11} are defined as:

$$\begin{cases} i \in I_{00}(w) := \{i : h_i/\sigma^2 \leq w\}, \\ i \in I_{10}(w, v) := \{i : g_i w < h_i g_i/\sigma^2 \leq h_i v + g_i w\}, \\ i \in I_{11}(w, v) := \{i : h_i v + g_i w < h_i g_i/\sigma^2\}. \end{cases} \tag{6.13}$$

As can be seen, the optimal power allocation for the user given in (6.11) and the optimal power allocation for the adversary given in (6.12) are mutually dependent on each other, but a numerical solution can be obtained using an iterative algorithm as provided in [29].

6.2 Problem Formulation

6.2.1 System Description

We consider a multiband communication system,[1] where there is *an authorized user* and *a jammer* who are operating against each other, without knowledge on the strategy applied by its opponent. Assuming that both the authorized user and the jammer can choose to operate over all or part of the N_c frequency bands or subchannels (not necessarily being consecutive), each of which has a bandwidth $\frac{B}{N_c}$ Hz. We start with the AWGN channel model, where all the subchannels have equal noise power, and then extend to the frequency selective fading scenario. In the AWGN case, assuming the total noise power over the entire spectrum is P_N, then the noise power corresponding to each subchannel is $\frac{P_N}{N_c}$. We assume the jamming is Gaussian over each jammed subchannel, because Gaussian jamming is the worst jamming when the jammer has no knowledge of the authorized transmission [13]. In the following, let P_s denote the total signal power for the authorized user and P_J the total jamming power.

The authorized user is always trying to maximize its capacity under jamming by applying an optimal strategy on subchannel selection (either all or part) and power allocation. On the other hand, the jammer tries to find an optimal strategy that can minimize the capacity of the authorized user. Here, we consider the case where both the authorized user and the jammer use mixed strategies. It is assumed that both the authorized user and the jammer can adjust their subchannel selection and power allocation swiftly and randomly, such that neither of them has sufficient time to learn and react in real time before its opponent switches to new subchannels and/or power levels. In other words, when the authorized user and the jammer apply their own resource allocation strategy, they have no knowledge of the selected subchannels and power levels of their opponent.

[1] We assume multiband communications, but the derivation here is readily applicable to multicarrier communications (e.g., OFDM), if the authorized user and the jammer apply the same transceiver structure.

6.2.2 *Strategy Spaces for the Authorized User and the Jammer*

Each mixed strategy applied by the authorized user is determined by the number of activated subchannels, the subchannel selection process, and the power allocation process. More specifically, (1) the authorized user activates K_s ($1 \leq K_s \leq N_c$) out of N_c subchannels each time for information transmission. (2) The subchannel selection process is characterized using a binary indicator vector $\boldsymbol{\alpha} = [\alpha_1, \alpha_2, \ldots, \alpha_{N_c}]$, where each random variable $\alpha_m = 1$ or 0 indicates whether the mth subchannel is selected or not, and $\sum_{m=1}^{N_c} \alpha_m = K_s$. Let $\boldsymbol{\omega}_s = [\omega_{s,1}, \omega_{s,2}, \ldots, \omega_{s,N_c}]$ be the corresponding probability vector, where $\omega_{s,m}$ denotes the probability that the mth subchannel is selected each time. That is, $\omega_{s,m} = Pr\{\alpha_m = 1\}$, and $\sum_{m=1}^{N_c} \omega_{s,m} = K_s$. (A simple strategy for selecting a particular number of subchannels based on a given subchannel selection probability vector, $\boldsymbol{\omega}_s$, is illustrated in Appendix C.) (3) For notation simplicity, the authorized user always specifies the indices of the selected K_s subchannels as $1, 2, \ldots, K_s$, following the order as they appear in the original spectrum, and performs power allocation over them. The power allocation process is characterized using a vector $\mathbf{P}_s = [P_{s,1}, P_{s,2}, \ldots, P_{s,K_s}]$, in which $P_{s,n}$ denotes the power allocated to the nth selected subchannel, and $\sum_{n=1}^{K_s} P_{s,n} = P_s$ is the power constraint. Let $\mathcal{W}_{s,K_s} = \{\boldsymbol{\omega}_s = [\omega_{s,1}, \omega_{s,2}, \ldots, \omega_{s,N_c}] \mid 0 \leq \omega_{s,m} \leq 1, \sum_{m=1}^{N_c} \omega_{s,m} = K_s\}$, and $\mathcal{P}_{s,K_s} = \{\mathbf{P}_s = [P_{s,1}, P_{s,2}, \ldots, P_{s,K_s}] \mid 0 < P_{s,n} \leq P_s, \sum_{n=1}^{K_s} P_{s,n} = P_s\}$. The strategy space for the authorized user can thus be defined as

$$\mathcal{X} = \{(K_s, \boldsymbol{\omega}_s, \mathbf{P}_s) \mid 1 \leq K_s \leq N_c, \boldsymbol{\omega}_s \in \mathcal{W}_{s,K_s}, \mathbf{P}_s \in \mathcal{P}_{s,K_s}\}. \tag{6.14}$$

The strategy space $\mathcal{X}$ covers all the possible subchannel utilization strategies as K_s varies from 1 to N_c. Here, a strategy $(K_s, \boldsymbol{\omega}_s, \mathbf{P}_s)$ with $K_s = 1$ and $\boldsymbol{\omega}_s = [\frac{1}{N_c}, \cdots, \frac{1}{N_c}]$ and $P_{s,1} = P_s$ corresponds to the conventional frequency hopping (FH) system, while a strategy $(K_s, \boldsymbol{\omega}_s, \mathbf{P}_s)$ with $K_s = N_c$, $\boldsymbol{\omega}_s = [1, \cdots, 1]$ and $P_{s,n} = \frac{P_s}{N_c}$, $\forall n$ would result in a full-band transmission with uniform power allocation.

Similarly, the jammer jams K_J ($1 \leq K_J \leq N_c$) out of N_c subchannels each time following a binary indicator vector $\boldsymbol{\beta} = [\beta_1, \beta_2, \ldots, \beta_{N_c}]$ with $\sum_{m=1}^{N_c} \beta_m = K_J$. The subchannel selection process is characterized using a probability vector $\boldsymbol{\omega}_J = [\omega_{J,1}, \omega_{J,2}, \ldots, \omega_{J,N_c}]$, where $\omega_{J,m} = Pr\{\beta_m = 1\}$ and $\sum_{m=1}^{N_c} \omega_{J,m} = K_J$. Then the jammer specifies the indices of the K_J jammed subchannels as $1, 2, \ldots, K_J$ in the same manner as the authorized user and performs power allocation over them using a power allocation vector $\mathbf{P}_J = [P_{J,1}, P_{J,2}, \ldots, P_{J,K_J}]$ constrained by $\sum_{n=1}^{K_J} P_{J,n} = P_J$. Let $\mathcal{W}_{J,K_J} = \{\boldsymbol{\omega}_J = [\omega_{J,1}, \omega_{J,2}, \ldots, \omega_{J,N_c}] \mid 0 \leq \omega_{J,m} \leq 1, \sum_{m=1}^{N_c} \omega_{J,m} = K_J\}$ and $\mathcal{P}_{J,K_J} = \{\mathbf{P}_J = [P_{J,1}, P_{J,2}, \ldots, P_{J,K_J}] \mid 0 < P_{J,n} \leq P_J, \sum_{n=1}^{K_J} P_{J,n} = P_J\}$; the strategy space for the jammer can thus be defined as

$$\mathcal{Y} = \{(K_J, \boldsymbol{\omega}_J, \mathbf{P}_J) \mid 1 \leq K_J \leq N_c, \boldsymbol{\omega}_J \in \mathcal{W}_{J,K_J}, \mathbf{P}_J \in \mathcal{P}_{J,K_J}\}. \tag{6.15}$$

6.2.3 The Minimax Problem in the Zero-Sum Game Between the Authorized User and the Jammer

From a game theoretic perspective, the strategic decision-making of the authorized user and the jammer can be modeled as a two-party zero-sum game [30], which is characterized by a triplet $(\mathcal{X}, \mathcal{Y}, C)$, where

1. $\mathcal{X}$ is the strategy space of the authorized user;
2. $\mathcal{Y}$ is the strategy space of the jammer;
3. C is a real-valued payoff function defined on $\mathcal{X} \times \mathcal{Y}$.

The interpretation is as follows. Let (x, y) denote the strategy pair, in which $x \in \mathcal{X}$ and $y \in \mathcal{Y}$ are the strategies applied by the authorized user and the jammer, respectively. Note that both x and y are mixed strategies. The payoff function $C(x, y)$ is therefore defined as the *ergodic* (i.e., expected or average) capacity of the authorized user choosing a strategy $x \in \mathcal{X}$ in the presence of the jammer choosing a strategy $y \in \mathcal{Y}$. In other words, $C(x, y)$ is the amount that the authorized user wins and simultaneously the jammer loses in the game with a strategy pair (x, y).

Assuming that with strategy pair (x, y), the authorized user and the jammer activate K_s and K_J channels, respectively. Define $\mathcal{A}_{K_s} = \{\boldsymbol{\alpha} = [\alpha_1, \alpha_2, \ldots, \alpha_{N_c}] \mid \alpha_m \in \{0, 1\}, \sum_{m=1}^{N_c} \alpha_m = K_s\}$ and $\mathcal{B}_{K_J} = \{\boldsymbol{\beta} = [\beta_1, \beta_2, \ldots, \beta_{N_c}] \mid \beta_m \in \{0, 1\}, \sum_{m=1}^{N_c} \beta_m = K_J\}$. Let $p(\boldsymbol{\alpha}|x)$ denote the probability that the subchannels selected by the authorized user follow the indicator vector $\boldsymbol{\alpha}$ given that the strategy $x \in \mathcal{X}$ is applied and $p(\boldsymbol{\beta}|y)$ the probability that the subchannels selected by the jammer follow the indicator vector $\boldsymbol{\beta}$ given that the strategy $y \in \mathcal{Y}$ is applied. Let $T_{s,m}$ and $T_{J,m}$ be the power allocated to the mth subchannel by the authorized user and the jammer, respectively, which are determined by

$$T_{s,m} = \begin{cases} P_{s,g_m}, & \alpha_m = 1, \\ 0, & \alpha_m = 0, \end{cases} \qquad T_{J,m} = \begin{cases} P_{J,q_m}, & \beta_m = 1, \\ 0, & \beta_m = 0, \end{cases} \tag{6.16}$$

where $g_m = \sum_{i=1}^{m} \alpha_i$ is the new index of subchannel m specified by the authorized user in the K_s selected subchannels if it is activated by the authorized user ($\alpha_m = 1$), and $q_m = \sum_{i=1}^{m} \beta_i$ is the new index of subchannel m specified by the jammer in the K_J jammed subchannels if it is activated by the jammer ($\beta_m = 1$). Note that (i) the subchannel selection processes used by the authorized user and the jammer are independent of each other and (ii) for each strategy pair (x, y), the subchannel selection choices ($\boldsymbol{\alpha}$ and $\boldsymbol{\beta}$) are not unique for both the authorized user and the jammer. Thus, the ergodic capacity of the authorized user in the game with a strategy pair (x, y) can be calculated as

$$C(x, y) = \sum_{\boldsymbol{\alpha} \in \mathcal{A}_{K_s}} \sum_{\boldsymbol{\beta} \in \mathcal{B}_{K_J}} p(\boldsymbol{\alpha}|x) p(\boldsymbol{\beta}|y) \sum_{m=1}^{N_c} \frac{B}{N_c} \log_2 \left(1 + \frac{T_{s,m}}{T_{J,m} + P_N/N_c} \right). \tag{6.17}$$

In the jamming-free case, the traditional Shannon channel capacity is obtained by maximizing the mutual information with respect to the distribution of the user signal. When jamming is around, the user still wants to maximize its capacity, while the jammer tries to minimize the user's capacity. This is why the "minimax" capacity was introduced in literature [31–33], for which the mutual information is *maximized* with respect to the distribution of the user signal and meanwhile *minimized* with respect to the distribution of the jamming. Now, in addition to optimizing the signal distribution, both the authorized user and the jammer can also choose which subchannels to use or activate and how much power to be allocated to each activated subchannel. That is, the minimax capacity is obtained through the optimization with respect to the strategies from both the authorized user and the jammer sides, in addition to the signal distributions.

Based on the definitions and reasoning above, the minimax capacity of the authorized user is defined as

$$C(x^*, y^*) = \max_{x\in\mathcal{X}} \min_{y\in\mathcal{Y}} C(x, y) = \min_{y\in\mathcal{Y}} \max_{x\in\mathcal{X}} C(x, y). \tag{6.18}$$

It can be seen from (6.18) that the authorized user tries to choose an optimal transmission strategy $x^* \in \mathcal{X}$ to maximize its capacity, while the jammer tries to minimize it by choosing an optimal jamming strategy $y^* \in \mathcal{Y}$. The capacity $C(x^*, y^*)$ in (6.18) can be achieved when a saddle-point strategy pair (x^*, y^*) is chosen, which is characterized by [3, 16]

$$C(x, y^*) \le C(x^*, y^*) \le C(x^*, y), \quad \forall x \in \mathcal{X}, y \in \mathcal{Y}. \tag{6.19}$$

This implies that with strategy x^*, the minimal capacity that can be achieved by the authorized user is $C(x^*, y^*)$, no matter which strategy is applied by the jammer; on the other hand, if the jammer applies strategy y^*, the maximal capacity that can be achieved by the authorized user is also $C(x^*, y^*)$, no matter which strategy is applied by the authorized user. As a result, to find the optimal transmission strategy and the worst jamming strategy under the power constraints P_s and P_J, we need to find the saddle-point strategy pair (x^*, y^*).

6.3 Multiband Communications Under Jamming Over AWGN Channels

Recall that K_s denotes the number of subchannels activated by the authorized user and K_J the number of subchannels interfered by the jammer. In this section, we derive the saddle-point strategy pair (x^*, y^*) in two steps: (1) For any fixed K_s and K_J with $1 \le K_s, K_J \le N_c$, calculate the corresponding minimax capacity and denote it by $\widetilde{C}(K_s, K_J)$. Let $K_s = 1, 2, \ldots, N_c$ and $K_J = 1, 2, \ldots, N_c$, we can obtain an $N_c \times N_c$ payoff matrix $\widetilde{\mathbf{C}}$. (2) For the derived payoff matrix $\widetilde{\mathbf{C}}$, locate its saddle point, and then the minimax capacity of the authorized user in (6.18) can be calculated accordingly.

6.3.1 The Minimax Problem for Fixed K_s and K_J

With fixed K_s and K_J, the strategy space for the authorized user becomes $\widetilde{\mathcal{X}}_{K_s} = \{(K_s, \boldsymbol{\omega}_s, \mathbf{P}_s) \mid K_s \text{ Fixed}, \boldsymbol{\omega}_s \in \mathcal{W}_{s,K_s}, \mathbf{P}_s \in \mathcal{P}_{s,K_s}\} \subset \mathcal{X}$, and similarly the strategy space for the jammer becomes $\widetilde{\mathcal{Y}}_{K_J} = \{(K_J, \boldsymbol{\omega}_J, \mathbf{P}_J) \mid K_J \text{ Fixed}, \boldsymbol{\omega}_J \in \mathcal{W}_{J,K_J}, \mathbf{P}_J \in \mathcal{P}_{J,K_J}\} \subset \mathcal{Y}$. It should be noted that the user-activated subchannels and the jammed subchannels may vary from time to time, although the total number of the user-activated or jammed subchannels is fixed.

We first present two lemmas on the concavity/convexity property of two real-valued functions that will be used afterward. More information on concavity and convexity can be found in [34].

Lemma 6.1 *For any $v \geq 0$ and $a > 0$, the real-valued function, $f(v) = \log_2(1 + \frac{v}{a})$, is concave.*

Proof The second-order derivative, $f''(v) = -\frac{1}{\ln 2}\frac{1}{(v+a)^2} < 0$, for any $v \geq 0$ and $a > 0$. ■

Lemma 6.2 *For any $v \geq 0$, $a > 0$, and $b > 0$, the real-valued function, $f(v) = \log_2(1 + \frac{a}{v+b})$, is convex.*

Proof The second-order derivative, $f''(v) = \frac{a}{\ln 2}\frac{(2v+a+2b)}{(v+a)^2(v+a+b)^2} > 0$, for any $v \geq 0$, $a > 0$, and $b > 0$. ■

The solution[2] to the minimax problem for fixed K_s and K_J is given in Theorem 6.3.

Theorem 6.3 *Let K_s be the number of subchannels activated by the authorized user and K_J the number of subchannels interfered by the jammer. For any fixed (K_s, K_J) pair, the saddle point of $C(x, y)$ under the power constraints P_s and P_J for $x \in \widetilde{\mathcal{X}}_{K_s}$ and $y \in \widetilde{\mathcal{Y}}_{K_J}$ is reached when both authorized user and the jammer choose to apply uniform subchannel selection and uniform power allocation strategy. That is, for fixed K_s and K_J, the saddle-point strategy pair $(\widetilde{x}^*, \widetilde{y}^*)$ that satisfies*

$$C(\widetilde{x}, \widetilde{y}^*) \leq C(\widetilde{x}^*, \widetilde{y}^*) \leq C(\widetilde{x}^*, \widetilde{y}), \quad \forall \widetilde{x} \in \widetilde{\mathcal{X}}_{K_s}, \widetilde{y} \in \widetilde{\mathcal{Y}}_{K_J}, \tag{6.20}$$

is given by $\widetilde{x}^* = (K_s, \boldsymbol{\omega}_s^*, \boldsymbol{P}_s^*)$ with

$$\begin{cases} \omega_{s,m}^* = K_s/N_c, & m = 1, 2, \ldots, N_c, \\ P_{s,n}^* = P_s/K_s, & n = 1, 2, \ldots, K_s, \end{cases} \tag{6.21}$$

and $\widetilde{y}^* = (K_J, \boldsymbol{\omega}_J^*, \boldsymbol{P}_J^*)$ with

$$\begin{cases} \omega_{J,m}^* = K_J/N_c, & m = 1, 2, \ldots, N_c, \\ P_{J,n}^* = P_J/K_J, & n = 1, 2, \ldots, K_J. \end{cases} \tag{6.22}$$

[2] The uniqueness of the solution is discussed in Appendix D.

In this case, the minimax capacity of the authorized user is given by

$$\begin{aligned}\widetilde{C}(K_s, K_J) =& K_s \frac{K_J}{N_c}\frac{B}{N_c}\log_2\left(1+\frac{P_s/K_s}{P_J/K_J+P_N/N_c}\right)\\&+K_s\left(1-\frac{K_J}{N_c}\right)\frac{B}{N_c}\log_2\left(1+\frac{P_s/K_s}{P_N/N_c}\right).\end{aligned} \tag{6.23}$$

Proof (1) We first prove that the $(\widetilde{x}^*, \widetilde{y}^*)$ pair defined in (6.21) and (6.22) satisfies the left part of (6.20), $C(\widetilde{x}, \widetilde{y}^*) \leq C(\widetilde{x}^*, \widetilde{y}^*)$. Assume the jammer applies the strategy $\widetilde{y}^*$, which means uniform subchannel selection and uniform power allocation as indicated in (6.22). For the authorized user who applies an arbitrary strategy $\widetilde{x} \in \widetilde{\mathcal{X}}_{K_s}$, we specified the indices of the activated K_s subchannels as $n = 1, 2, \ldots, K_s$. With any subchannel selection probability vector $\boldsymbol{\omega}_s \in \mathcal{W}_{s,K_s}$, for each subchannel activated by the authorized user, the probability that it is jammed is always $\frac{K_J}{N_c}$, since the jammer jams each subchannel with a uniform probability $\omega^*_{J,m} = \frac{K_J}{N_c}$, for any $m = 1, 2, \ldots, N_c$. Accordingly, the probability that each subchannel is not jammed is $1 - \frac{K_J}{N_c}$.

Considering all the subchannels activated by the authorized user, when the authorized user applies an arbitrary strategy $\widetilde{x} \in \widetilde{\mathcal{X}}_{K_s}$, and the jammer applies strategy $\widetilde{y}^*$, the ergodic capacity can be calculated as the weighted average of the capacity under jamming and the capacity in the jamming-free case:

$$\begin{aligned}C(\widetilde{x}, \widetilde{y}^*) =& \sum_{n=1}^{K_s}\left[\frac{K_J}{N_c}\frac{B}{N_c}\log_2\left(1+\frac{P_{s,n}}{P_J/K_J+P_N/N_c}\right)\right.\\&\left.+\left(1-\frac{K_J}{N_c}\right)\frac{B}{N_c}\log_2\left(1+\frac{P_{s,n}}{P_N/N_c}\right)\right]\\=&\frac{K_J}{N_c}\frac{B}{N_c}\sum_{n=1}^{K_s}\log_2\left(1+\frac{P_{s,n}}{P_J/K_J+P_N/N_c}\right)\\&+\left(1-\frac{K_J}{N_c}\right)\frac{B}{N_c}\sum_{n=1}^{K_s}\log_2\left(1+\frac{P_{s,n}}{P_N/N_c}\right).\end{aligned} \tag{6.24}$$

Note that $\sum_{n=1}^{K_s} P_{s,n} = P_s$, and applying the concavity property proved in Lemma 6.1, we have

$$\begin{aligned}C(\widetilde{x}, \widetilde{y}^*) \leq& K_s\frac{K_J}{N_c}\frac{B}{N_c}\log_2\left(1+\frac{P_s/K_s}{P_J/K_J+P_N/N_c}\right)\\&+K_s\left(1-\frac{K_J}{N_c}\right)\frac{B}{N_c}\log_2\left(1+\frac{P_s/K_s}{P_N/N_c}\right)\\=&C(\widetilde{x}^*, \widetilde{y}^*),\end{aligned} \tag{6.25}$$

where the equality holds if and only if $P_{s,n} = \frac{P_s}{K_s}, \forall n$.

(2) Proof of the right part of (6.20), $C(\widetilde{x}^*, \widetilde{y}^*) \leq C(\widetilde{x}^*, \widetilde{y})$. In this part of the proof, we will show that applying uniform subchannel selection and uniform power allocation strategy $\widetilde{x}^*$ at the authorized user side guarantees a lower bound on its capacity, no matter what strategy is applied by the jammer. Assume the authorized user applies the strategy $\widetilde{x}^*$ as indicated in (6.21). For the jammer who applies an arbitrary strategy $\widetilde{y} \in \widetilde{\mathcal{Y}}_{K_J}$, we specified the indices of the jammed K_J subchannels as $n = 1, 2, \ldots, K_J$. With any subchannel selection probability vector $\boldsymbol{\omega}_J \in \mathcal{W}_{J,K_J}$, for each jammed or jamming-free subchannel, the probability that it serves as a subchannel activated by the authorized user is always $\frac{K_s}{N_c}$. Hence, the average number[3] of jammed subchannels which are also activated by the authorized user is $\frac{K_J K_s}{N_c}$, and the average number of jamming-free subchannels which are activated by the authorized user would be $(N_c - K_J)\frac{K_s}{N_c} = K_s(1 - \frac{K_J}{N_c})$.

Considering both the jammed and jamming-free subchannels, when the jammer applies an arbitrary strategy $\widetilde{y} \in \widetilde{\mathcal{Y}}_{K_J}$, and the authorized user applies strategy $\widetilde{x}^*$, the ergodic capacity can be calculated as

$$
\begin{aligned}
C(\widetilde{x}^*, \widetilde{y}) = &\sum_{n=1}^{K_J} \frac{K_s}{N_c}\frac{B}{N_c} \log_2\left(1 + \frac{P_s/K_s}{P_{J,n} + P_N/N_c}\right) \\
&+ K_s\left(1 - \frac{K_J}{N_c}\right)\frac{B}{N_c}\log_2\left(1 + \frac{P_s/K_s}{P_N/N_c}\right).
\end{aligned} \tag{6.26}
$$

Note that $\sum_{n=1}^{K_J} P_{J,n} = P_J$, and applying the convexity property proved in Lemma 6.2, we have

$$
\begin{aligned}
C(\widetilde{x}^*, \widetilde{y}) \geq &K_s \frac{K_J}{N_c}\frac{B}{N_c} \log_2\left(1 + \frac{P_s/K_s}{P_J/K_J + P_N/N_c}\right) \\
&+ K_s\left(1 - \frac{K_J}{N_c}\right)\frac{B}{N_c}\log_2\left(1 + \frac{P_s/K_s}{P_N/N_c}\right) \\
= &C(\widetilde{x}^*, \widetilde{y}^*),
\end{aligned} \tag{6.27}
$$

where the equality holds if and only if $P_{J,n} = \frac{P_J}{K_J}, \forall n$. ■

6.3.2 *Capacity Optimization Over K_s and K_J*

In Sect. 6.3.1, we derived the closed-form minimax capacity of the authorized user for fixed K_s and K_J. Considering all possible K_s and K_J, we would have an $N_c \times N_c$

[3] The ensemble average might not be an integer. Nevertheless, the capacity calculation would still be accurate from a statistical perspective.

matrix $\widetilde{\mathbf{C}}$, in which $\widetilde{C}(K_s, K_J)$ is the minimax capacity of the authorized user for fixed K_s and K_J, as indicated in (6.23). Now finding the minimax capacity in (6.18) can be reduced to finding the saddle point of the matrix $\widetilde{\mathbf{C}}$, that is, the entry $\widetilde{C}(i, j)$, which is simultaneously the minimum of the ith row and the maximum of the jth column.

To locate the saddle point of matrix $\widetilde{\mathbf{C}}$, we need Lemma 6.3.

Lemma 6.3 *For the capacity function*

$$\begin{aligned}\widetilde{C}(K_s, K_J) =& K_s \frac{K_J}{N_c}\frac{B}{N_c}\log_2\left(1 + \frac{P_s/K_s}{P_J/K_J + P_N/N_c}\right)\\ &+ K_s\left(1 - \frac{K_J}{N_c}\right)\frac{B}{N_c}\log_2\left(1 + \frac{P_s/K_s}{P_N/N_c}\right),\end{aligned} \tag{6.28}$$

we have

$$\frac{\partial \widetilde{C}}{\partial K_s} > 0, \text{ for any } K_s = 1, 2, \ldots, N_c, \tag{6.29}$$

and

$$\frac{\partial \widetilde{C}}{\partial K_J} < 0, \text{ for any } K_J = 1, 2, \ldots, N_c. \tag{6.30}$$

Proof See Appendix E. ■

Following Lemma 6.3, we have the following theorem.

Theorem 6.4 *The saddle point of matrix $\widetilde{\mathbf{C}}$ is indexed by $(K_s^*, K_J^*) = (N_c, N_c)$. Equivalently, for all $1 \le K_s, K_J \le N_c$, we have*

$$\widetilde{C}(K_s, N_c) \le \widetilde{C}(N_c, N_c) \le \widetilde{C}(N_c, K_J). \tag{6.31}$$

Combining Theorems 6.3 and 6.4, we can obtain the saddle point to the original minimax problem in (6.18) over strategy spaces $\mathcal{X}$ and $\mathcal{Y}$. The result is summarized in Theorem 6.5.

Theorem 6.5 *Assume that an authorized user and a jammer are operating against each other over the same AWGN channel consisting of N_c subchannels. Either for the authorized user to maximize its capacity or for the jammer to minimize the capacity of the authorized user, the best strategy for both of them is to distribute the signal power or jamming power uniformly over all the N_c subchannels. In this case, the minimax capacity of the authorized user is given by*

$$C = B\log_2\left(1 + \frac{P_s}{P_J + P_N}\right), \tag{6.32}$$

where B is the bandwidth of the overall spectrum, P_N the noise power, and P_s and P_J the total power for the authorized user and the jammer, respectively.

Proof The proof follows directly from Theorems 6.3 and 6.4. The minimax capacity in (6.32) can be derived simply by substituting $K_s = K_J = N_c$ into (6.23). ■

6.4 Multiband Communications Under Jamming Over Frequency Selective Fading Channels

In this section, we investigate the optimal strategies for both the authorized user and the jammer in multiband communications under frequency selective fading channels.

6.4.1 The Minimax Problem for Fading Channels

Recall that the power allocation for the authorized user is characterized using the vector $\mathbf{P}_s = [P_{s,1}, P_{s,2}, \ldots, P_{s,N_c}]$, where $P_{s,i}$ denotes the power allocated to the ith subchannel, and $\sum_{i=1}^{N_c} P_{s,i} = P_s$ is the signal power constraint. Similarly, the power allocation vector for the jammer is $\mathbf{P}_J = [P_{J,1}, P_{J,2}, \ldots, P_{J,N_c}]$, and $\sum_{i=1}^{N_c} P_{J,i} = P_J$ is the jamming power constraint. As in the OFDM systems, here we assume that all the subchannels are narrowband and have flat magnitude spectrum. Let $\mathbf{H}_s = [H_{s,1}, H_{s,2}, \ldots, H_{s,N_c}]$ be the frequency domain channel response vector for the authorized user and $\mathbf{H}_J = [H_{J,1}, H_{J,2}, \ldots, H_{J,N_c}]$ the frequency domain channel response vector for the jammer, respectively. Under the settings specified above, the capacity of the authorized user can be calculated as

$$\begin{aligned} C(\mathbf{P}_s, \mathbf{P}_J) &= \sum_{i=1}^{N_c} \frac{B}{N_c} \log_2 \left(1 + \frac{|H_{s,i}|^2 P_{s,i}}{|H_{J,i}|^2 P_{J,i} + \sigma_n^2}\right) \\ &= \sum_{i=1}^{N_c} \frac{B}{N_c} \log_2 \left(1 + \frac{P_{s,i}}{\frac{|H_{J,i}|^2}{|H_{s,i}|^2} P_{J,i} + \sigma_{n,i}^2}\right), \end{aligned} \tag{6.33}$$

where $\sigma_n^2 = \frac{P_N}{N_c}$ is the original noise power for each subchannel, and $\sigma_{n,i}^2 = \frac{\sigma_n^2}{|H_{s,i}|^2}$.

Define $\mathcal{P}_s = \{\mathbf{P}_s = [P_{s,1}, P_{s,2}, \ldots, P_{s,N_c}] \mid 0 \le P_{s,i} \le P_s, \sum_{i=1}^{N_c} P_{s,i} = P_s\}$ and $\mathcal{P}_J = \{\mathbf{P}_J = [P_{J,1}, P_{J,2}, \ldots, P_{J,N_c}] \mid 0 \le P_{J,i} \le P_J, \sum_{i=1}^{N_c} P_{J,i} = P_J\}$. The minimax capacity of the authorized user is defined as

$$C(\mathbf{P}_s^*, \mathbf{P}_J^*) = \max_{\mathbf{P}_s^* \in \mathcal{P}_s} \min_{\mathbf{P}_J^* \in \mathcal{P}_J} C(\mathbf{P}_s, \mathbf{P}_J) = \min_{\mathbf{P}_J^* \in \mathcal{P}_J} \max_{\mathbf{P}_s^* \in \mathcal{P}_s} C(\mathbf{P}_s, \mathbf{P}_J). \tag{6.34}$$

As before, the authorized user tries to apply optimal signal power allocation $\mathbf{P}_s^* \in \mathcal{P}_s$ to maximize its capacity, while the jammer tries to minimize it by applying optimal jamming power allocation $\mathbf{P}_J^* \in \mathcal{P}_J$.

Theorem 6.6 *Assume that there are N_c available subchannels. Let $\boldsymbol{H}_s = [H_{s,1}, H_{s,2}, \ldots, H_{s,N_c}]$ and $\boldsymbol{H}_J = [H_{J,1}, H_{J,2}, \ldots, H_{J,N_c}]$ denote the frequency domain channel response vector for the authorized user and the jammer, respectively. Assuming zero-mean white Gaussian noise of variance σ_n^2 over the entire band, let $\boldsymbol{\sigma}_n^2 = [\sigma_{n,1}^2, \sigma_{n,2}^2, \ldots, \sigma_{n,N_c}^2]$, where $\sigma_{n,i}^2 = \frac{\sigma_n^2}{|H_{s,i}|^2}$. The optimal power allocation pair for the authorized user and the jammer under the signal power constraint $\sum_{i=1}^{N_c} P_{s,i}^* = P_s$ and the jamming power constraint $\sum_{i=1}^{N_c} P_{J,i}^* = P_J$, $(\boldsymbol{P}_s^*, \boldsymbol{P}_J^*)$, which satisfies*

$$C(\boldsymbol{P}_s, \boldsymbol{P}_J^*) \leq C(\boldsymbol{P}_s^*, \boldsymbol{P}_J^*) \leq C(\boldsymbol{P}_s^*, \boldsymbol{P}_J), \quad \forall \boldsymbol{P}_s \in \mathcal{P}_s, \boldsymbol{P}_J \in \mathcal{P}_J, \tag{6.35}$$

can be characterized by

$$\begin{cases} P_{J,i}^* = sgn(P_{s,i}^*)\left(c_1 - \dfrac{|H_{s,i}|^2}{|H_{J,i}|^2}\sigma_{n,i}^2\right)^+, \quad \forall i, & (6.36a) \\ P_{s,i}^* = \left(c_2 - \dfrac{|H_{J,i}|^2}{|H_{s,i}|^2}P_{J,i}^* - \sigma_{n,i}^2\right)^+, \quad \forall i, & (6.36b) \end{cases}$$

where $(x)^+ = \max\{0, x\}$, $sgn(\cdot)$ is the sign function, and c_1, c_2 are constants determined by the power constraints.

Proof (1) We first prove that the $(\mathbf{P}_s^*, \mathbf{P}_J^*)$ pair defined in (6.36) satisfies the left part of (6.35), $C(\mathbf{P}_s, \mathbf{P}_J^*) \leq C(\mathbf{P}_s^*, \mathbf{P}_J^*)$, $\forall \mathbf{P}_s \in \mathcal{P}_s$. With the jammer applying power allocation $\mathbf{P}_J^*$, the equivalent jamming power for the ith subchannel after fading and equalization would be $\frac{|H_{J,i}|^2}{|H_{s,i}|^2} P_{J,i}^*$, as shown in (6.33). Taking both the jamming and the noise into account, the overall interference and noise power level for the ith subchannel at the receiver would be $\frac{|H_{J,i}|^2}{|H_{s,i}|^2} P_{J,i}^* + \sigma_{n,i}^2$. As a result, the problem now turns to be the capacity maximization problem for multiband communications with nonuniform noise power levels. To this end, it is well known that the classical water-filling algorithm produces the best solution [35]. In this particular case, the water-filling solution for optimal signal power allocation would be

$$P_{s,i}^* = \left(c_2 - \frac{|H_{J,i}|^2}{|H_{s,i}|^2} P_{J,i}^* - \sigma_{n,i}^2\right)^+, \quad i = 1, 2, \ldots, N_c, \tag{6.37}$$

which maximizes the capacity of the authorized user, $C(\mathbf{P}_s^*, \mathbf{P}_J^*)$, while the jammer applying power allocation $\mathbf{P}_J^*$. Note that c_2 is a constant that should be chosen such that the power constraint for the authorized user is satisfied, i.e., $\sum_{i=1}^{N_c} P_{s,i}^* = P_s$.

(2) Proof of the right part, $C(\mathbf{P}_s^*, \mathbf{P}_J^*) \le C(\mathbf{P}_s^*, \mathbf{P}_J)$, $\forall \mathbf{P}_J \in \mathcal{P}_J$. To this end, we need to find the optimal jamming power allocation $\mathbf{P}_J^*$, which can minimize the capacity of the authorized user applying power allocation $\mathbf{P}_s^*$. Let $\gamma_i = \frac{|H_{J,i}|^2}{|H_{s,i}|^2}$, $\forall i$. With the authorized user applying power allocation $\mathbf{P}_s^*$, following (6.33), the optimization problem for jamming power allocation can be formulated as

$$\min_{\mathbf{P}_J \in \mathcal{P}_J} \sum_{i=1}^{N_c} \frac{B}{N_c} \log_2 \left(1 + \frac{P_{s,i}^*}{\gamma_i P_{J,i} + \sigma_{n,i}^2}\right); \tag{6.38a}$$

$$\text{s.t.} \sum_{i=1}^{N_c} P_{J,i} = P_J, \tag{6.38b}$$

$$P_{J,i} \ge 0, \ \forall i. \tag{6.38c}$$

Note that in this optimization problem, we have both equality and inequality constraints. Hence, we need to take the Karush-Kuhn-Tucker (KKT) approach [34], which generalizes the conventional method of Lagrange multipliers by allowing inequality constraints. As observed in (6.37), for $P_{s,i}^* > 0$, $P_{s,i}^* = c_2 - \frac{|H_{J,i}|^2}{|H_{s,i}|^2} P_{J,i}^* - \sigma_{n,i}^2$. In addition, the capacity of any subchannel with zero signal power (i.e., $P_{s,i}^* = 0$) is zero. Define

$$\begin{aligned} J(\mathbf{P}_J, \mathbf{u}, v) &= \sum_{i=1}^{N_c} \frac{B}{N_c} \log_2 \left(1 + \frac{P_{s,i}^*}{\gamma_i P_{J,i} + \sigma_{n,i}^2}\right) - u_i P_{J,i} + v \left(\sum_{i=1}^{N_c} P_{J,i} - P_J\right) \\ &= \sum_{i \in \{i | P_{s,i}^* > 0\}} \frac{B}{N_c} \log_2 \frac{c_2}{\gamma_i P_{J,i} + \sigma_{n,i}^2} - u_i P_{J,i} + v \left(\sum_{i=1}^{N_c} P_{J,i} - P_J\right), \end{aligned} \tag{6.39}$$

where $\mathbf{u} = [u_1, u_2, \ldots, u_{N_c}]$ and v are Lagrange multipliers that should satisfy the KKT conditions as below:

$$\frac{\partial J}{\partial P_{J,i}} = 0, \ \ u_i P_{J,i} = 0, \ \ u_i \ge 0, \ \ \forall i. \tag{6.40}$$

The first-order partial differentiation with respect to each $P_{J,i}$ can be calculated as

$$\frac{\partial J}{\partial P_{J,i}} = \begin{cases} -\frac{B}{N_c} \frac{1}{\ln 2} \frac{\gamma_i}{\gamma_i P_{J,i} + \sigma_{n,i}^2} - u_i + v, & P_{s,i}^* > 0, \\ -u_i + v, & P_{s,i}^* = 0. \end{cases} \tag{6.41}$$

For each subchannel with nonzero signal power (i.e., $P_{s,i}^* > 0$), applying the KKT conditions and eliminating u_i, we have

$$\begin{cases} v - \frac{B}{N_c} \frac{1}{\ln 2} \frac{\gamma_i}{\gamma_i P_{J,i} + \sigma_{n,i}^2} \geq 0, \\ P_{J,i} \left[v - \frac{B}{N_c} \frac{1}{\ln 2} \frac{\gamma_i}{\gamma_i P_{J,i} + \sigma_{n,i}^2} \right] = 0. \end{cases} \tag{6.42}$$

Solving (6.42), the optimal jamming power for the ith subchannel (with nonzero signal power) can be obtained as

$$P_{J,i}^* = \left(\frac{B}{N_c} \frac{1}{\ln 2} \frac{1}{v} - \frac{1}{\gamma_i} \sigma_{n,i}^2 \right)^+ . \tag{6.43}$$

Similarly, for each subchannel with zero signal power (i.e., $P_{s,i}^* = 0$), applying the KKT conditions and eliminating u_i, we have $v P_{J,i} = 0$. It is observed from (6.42) that $v > 0$, so the optimal jamming power for the ith subchannel (with zero signal power) is $P_{J,i}^* = 0$. Let $c_1 = \frac{B}{N_c} \frac{1}{\ln 2} \frac{1}{v}$, and replace γ_i with $\frac{|H_{J,i}|^2}{|H_{s,i}|^2}$, we can summarize the result as

$$P_{J,i}^* = \begin{cases} \left(c_1 - \frac{|H_{s,i}|^2}{|H_{J,i}|^2} \sigma_{n,i}^2 \right)^+, & P_{s,i}^* > 0, \\ 0, & P_{s,i}^* = 0, \end{cases} \tag{6.44}$$

where c_1 should be chosen such that the power constraint for the jammer is satisfied, i.e., $\sum_{i=1}^{N_c} P_{J,i}^* = P_J$. This is exactly the optimal jamming power allocation as expressed in (6.36a), which minimizes the capacity of the authorized user, $C(\mathbf{P}_s^*, \mathbf{P}_J^*)$, given that the authorized user applies power allocation $\mathbf{P}_s^*$. ■

6.4.2 Correlated Fading Channels: A Two-Step Water-Filling Algorithm

Theorem 6.6 characterizes the dynamic relationship between the optimal signal power allocation $\mathbf{P}_s^*$ and the optimal jamming power allocation $\mathbf{P}_J^*$. As shown in (6.36), due to the mutual dependency between $\mathbf{P}_s^*$ and $\mathbf{P}_J^*$, it is generally difficult to find an exact solution for them. However, in this subsection, we will show that if the channels of the authorized user and the jammer are correlated, the saddle point, $(\mathbf{P}_s^*, \mathbf{P}_J^*)$, can be calculated explicitly using a two-step water-filling algorithm.

Theorem 6.7 (A Two-Step Water-Filling Algorithm) *With the same conditions as in Theorem 6.6, the saddle point, which indicates the optimal signal power allocation and the optimal jamming power allocation, is given by*

$$\begin{cases} P^*_{J,i} = \left(c_1 - \frac{|H_{s,i}|^2}{|H_{J,i}|^2} \sigma^2_{n,i} \right)^+, & \forall i, \quad (6.45a) \\ P^*_{s,i} = \left(c_2 - \frac{|H_{J,i}|^2}{|H_{s,i}|^2} P^*_{J,i} - \sigma^2_{n,i} \right)^+, & \forall i, \quad (6.45b) \end{cases}$$

as long as

$$|H_{J,i}|^2 \leq \frac{\sigma_n^2}{c_1} \text{ or } \frac{|H_{J,i}|^2}{|H_{s,i}|^2} < \frac{c_2}{c_1}, \ \forall i, \tag{6.46}$$

where $(x)^+ = \max\{0, x\}$ *and* c_1 *and* c_2 *are constants that should be chosen such that the power constraints are satisfied, i.e.,* $\sum_{i=1}^{N_c} P^*_{s,i} = P_s$ *and* $\sum_{i=1}^{N_c} P^*_{J,i} = P_J$.

Proof The basic idea here is that given zero signal power for a particular subchannel, it is apparently not necessary to allocate positive jamming power in that subchannel; at the same time, over all the subchannels with nonzero signal power, the optimal jamming power allocation can be formed using the water-filling algorithm. We start by applying the water-filling algorithm over all subchannels:

$$P^*_{J,i} = \left(c_1 - \frac{|H_{s,i}|^2}{|H_{J,i}|^2} \sigma^2_{n,i} \right)^+, \ i = 1, 2, \ldots, N_c. \tag{6.47}$$

For the optimality of (6.47), we further need to ensure that $P^*_{J,i} = 0$, whenever $P^*_{s,i} = 0$.

As can be seen, a violation occurs ($P^*_{J,i} > 0$ and $P^*_{s,i} = 0$), if and only if for some subchannel indexed by i,

$$\begin{cases} P^*_{J,i} = c_1 - \frac{|H_{s,i}|^2}{|H_{J,i}|^2} \sigma^2_{n,i} > 0, \\ c_2 - \frac{|H_{J,i}|^2}{|H_{s,i}|^2} P^*_{J,i} - \sigma^2_{n,i} \leq 0, \end{cases} \tag{6.48}$$

which yields

$$|H_{J,i}|^2 > \frac{\sigma_n^2}{c_1} \text{ and } \frac{|H_{J,i}|^2}{|H_{s,i}|^2} \geq \frac{c_2}{c_1}. \tag{6.49}$$

Note that $\sigma^2_{n,i} = \frac{\sigma_n^2}{|H_{s,i}|^2}$. Hence, the conditions characterized in (6.46) ensure that no violation occurs, and therefore the saddle point calculated by (6.45) is valid for both capacity maximization by the authorized user and capacity minimization by the jammer. ■

In the following, we consider a special case where the channels corresponding to the authorized user and the jammer are *relatively flat* with respect to each other, that is, their magnitude spectrum is proportional to each other, i.e., $\frac{|H_{J,i}|^2}{|H_{s,i}|^2} = \gamma,\ \forall i$. As will be shown in Corollary 6.1, when the user channel and the jammer channel are relatively flat with respect to each other, the conditions in (6.46) are always satisfied, and the saddle-point calculation can be simplified accordingly.

Corollary 6.1 *With the same conditions as in Theorem 6.6, if the magnitude spectrum of the channels for the authorized user and the jammer is proportional to each other, i.e., $\frac{|H_{J,i}|^2}{|H_{s,i}|^2} = \gamma,\ \forall i$, the saddle point, which indicates the optimal signal power allocation and the optimal jamming power allocation, can be calculated as*

$$\begin{cases} P_{J,i}^* = \left(c_1 - \frac{1}{\gamma}\sigma_{n,i}^2\right)^+, & \forall i, \\ P_{s,i}^* = \left(c_2 - \gamma P_{J,i}^* - \sigma_{n,i}^2\right)^+, & \forall i, \end{cases} \tag{6.50}$$

where $(x)^+ = \max\{0, x\}$ and c_1 and c_2 are constants that should be chosen such that the power constraints are satisfied, i.e., $\sum_{i=1}^{N_c} P_{s,i}^* = P_s$ and $\sum_{i=1}^{N_c} P_{J,i}^* = P_J$.

Proof Note that with $\frac{|H_{J,i}|^2}{|H_{s,i}|^2} = \gamma,\ \forall i$, (6.45) reduces to (6.50). Following Theorem 6.7, we only need to show that the conditions specified in (6.46) are satisfied.

First, we show that the constants c_1, c_2 resulted from (6.50) and the power constraints always satisfy $\frac{c_2}{c_1} > \gamma$. This is proved by contradiction as follows. Suppose $\frac{c_2}{c_1} \leq \gamma$. Following (6.50), for any $i = 1, 2, \ldots, N_c$, $P_{J,i}^* \geq c_1 - \frac{1}{\gamma}\sigma_{n,i}^2$. Thus, $c_2 - \gamma P_{J,i}^* - \sigma_{n,i}^2 \leq c_2 - \gamma(c_1 - \frac{1}{\gamma}\sigma_{n,i}^2) - \sigma_{n,i}^2 = c_2 - \gamma c_1 \leq 0$. This implies that for all subchannels, we always have $P_{s,i}^* = \left(c_2 - \gamma P_{J,i}^* - \sigma_{n,i}^2\right)^+ = 0$, which contradicts with the power constraint that $\sum_{i=1}^{N_c} P_{s,i}^* = P_s$. As a result, we must have $\frac{c_2}{c_1} > \gamma$.

It then follows that for any subchannel, we always have $\frac{|H_{J,i}|^2}{|H_{s,i}|^2} = \gamma < \frac{c_2}{c_1}$. This ensures that the conditions specified in (6.46) are always satisfied. Hence, the solution calculated by (6.50) must be a valid saddle point. ■

Furthermore, if the magnitude spectrum of channels for the authorized user and the jammer is equal to each other, i.e., $\frac{|H_{J,i}|^2}{|H_{s,i}|^2} = \gamma = 1,\ \forall i$, the two-step water-filling algorithm in (6.50) can be graphically illustrated in Fig. 6.1, where the saddle point can simply be obtained by filling all the signal power after filling all the jamming power into a tank with given noise power levels. We would like to point out that under AWGN channels, the noise power levels are flat; hence, the water-filling process here would result in uniform power allocation for both the jammer and the authorized user, which echoes the results in Sect. 6.3.

Discussions Theorem 6.7 provides an efficient two-step water-filling algorithm to calculate the saddle point of the minimax problem. This algorithm guarantees a valid

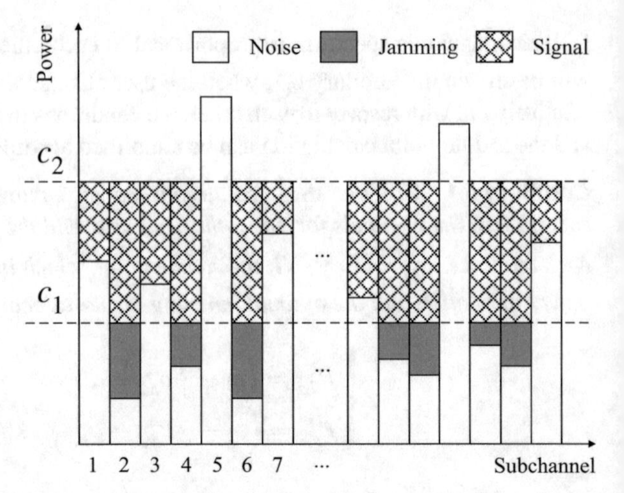

Fig. 6.1 Water filling under jamming with equal channel magnitude spectrum for the authorized user and the jammer (i.e., $\frac{|H_{J,i}|^2}{|H_{s,i}|^2} = \gamma = 1,\ \forall i$). (© [2016] IEEE. Reprinted, with permission, from Ref. [36])

saddle point under certain conditions as illustrated in (6.46). Corollary 6.1 further shows a sufficient (but may not be necessary) condition for (6.46) being satisfied: the channels for the authorized user and the jammer are relatively flat with respect to each other, i.e., their magnitude spectrum is proportional to each other. From the arbitrarily varying channel (AVC) [6, 7] point of view, the correlation between the user channel and the jamming channel can be regarded as an indicator of possible symmetricity between the user and the jammer. In the case that the user channel and the jammer channel are not relatively flat with respect to each other, as shown in Sect. 6.5.2, as long as the cross correlation between the two channels is reasonably high, we found that the algorithm in Theorem 6.7 can still provide a much better solution than uniform power allocation.

6.4.3 Arbitrary Fading Channels: An Iterative Water-Filling Algorithm

The two-step water-filling algorithm in Theorem 6.7 is a very efficient solution for correlated fading channels. However, if the channels of the authorized user and the jammer are not correlated, the algorithm needs to be extended. Motivated by [37], in this subsection, we will present an iterative water-filling algorithm, which is able to find a numerical solution to the saddle point for arbitrary fading channels.

We first begin with the two-step water-filling algorithm in Theorem 6.7, since it is a good starting point with possibly only a few violations against (6.46). We can then try to remove or at least alleviate the violations identified. Recall that in the two-step water-filling algorithm, we first allocate the jamming power by

$$P_{J,i}^* = \left(c_1 - \frac{|H_{s,i}|^2}{|H_{J,i}|^2} \sigma_{n,i}^2 \right)^+, \quad \forall i, \tag{6.51}$$

which is equivalent to

$$\frac{|H_{J,i}|^2}{|H_{s,i}|^2} P^*_{J,i} = \left(\frac{|H_{J,i}|^2}{|H_{s,i}|^2} c_1 - \sigma^2_{n,i} \right)^+, \quad \forall i. \tag{6.52}$$

The physical meaning of (6.52) is that for each subchannel i with positive jamming power allocation (i.e., $P^*_{J,i} > 0$), the effective jamming power, $\frac{|H_{J,i}|^2}{|H_{s,i}|^2} P^*_{J,i}$, plus the noise power level, $\sigma^2_{n,i}$, should be $\frac{|H_{J,i}|^2}{|H_{s,i}|^2} c_1$. However, if $\frac{|H_{J,i}|^2}{|H_{s,i}|^2} c_1 > c_2$, the allocated jamming power for this subchannel would be *more than necessary*. The underlying argument is that to prevent or discourage the transmission of the authorized user in a particular subchannel, it would be good enough to make sure the water level after jamming power allocation, $\frac{|H_{J,i}|^2}{|H_{s,i}|^2} P^*_{J,i} + \sigma^2_{n,i}$, reaches c_2. In this case, according to (6.45b), the authorized user would have already been discouraged from allocating any power in this subchannel. Hence, any jamming power that results into a water level higher than c_2 would be more than necessary. Based on the reasoning above, (6.52) should be revised to

$$\frac{|H_{J,i}|^2}{|H_{s,i}|^2} P^*_{J,i} = \left[\min \left(\frac{|H_{J,i}|^2}{|H_{s,i}|^2} c_1, c_2 \right) - \sigma^2_{n,i} \right]^+, \quad \forall i, \tag{6.53}$$

which is equivalent to

$$P^*_{J,i} = \left[\min \left(c_1, \frac{|H_{s,i}|^2}{|H_{J,i}|^2} c_2 \right) - \frac{|H_{s,i}|^2}{|H_{J,i}|^2} \sigma^2_{n,i} \right]^+, \quad \forall i. \tag{6.54}$$

Replacing the jamming power allocation in Theorem 6.7 by (6.54), we have

$$\begin{cases} P^*_{J,i} = \left[\min(c_1, \frac{|H_{s,i}|^2}{|H_{J,i}|^2} c_2) - \frac{|H_{s,i}|^2}{|H_{J,i}|^2} \sigma^2_{n,i} \right]^+, \quad \forall i, & (6.55a) \\ P^*_{s,i} = \left(c_2 - \frac{|H_{J,i}|^2}{|H_{s,i}|^2} P^*_{J,i} - \sigma^2_{n,i} \right)^+, \quad \forall i. & (6.55b) \end{cases}$$

Then we can approximate the optimal power allocation pair by alternatively running (6.55a) and (6.55b) until it converges. Following this idea, we present an iterative water-filling algorithm, which is summarized in Table 6.1.

It should be noted that for the jamming power allocation in step 3, c_2 is known from previous calculations. As a result, for each water-filling step throughout the algorithm, there is strictly only one constant unknown, and it can be determined by an efficient binary search algorithm [38]. The convergence analysis of the iterative water-filling algorithm can be found in Appendix F.

Table 6.1 The iterative water-filling algorithm. (© [2016] IEEE. Reprinted, with permission, from Ref. [36])

Step 1. Run the two-step water-filling algorithm once:

1. Allocate jamming power by
$P_{J,i}^* = \left(c_1 - \frac{|H_{s,i}|^2}{|H_{J,i}|^2}\sigma_{n,i}^2\right)^+, \quad \forall i;$
2. Allocate user signal power by
$P_{s,i}^* = \left(c_2 - \frac{|H_{J,i}|^2}{|H_{s,i}|^2}P_{J,i}^* - \sigma_{n,i}^2\right)^+, \quad \forall i.$

Step 2. Exit if no violations, i.e., $|H_{J,i}|^2 \leq \frac{\sigma_n^2}{c_1}$ or $\frac{|H_{J,i}|^2}{|H_{s,i}|^2} < \frac{c_2}{c_1}$, $\forall i$;
Step 3. Repeat the following water-filling steps until convergence:

1. Allocate jamming power by
$P_{J,i}^* = \left[\min\left(c_1, \frac{|H_{s,i}|^2}{|H_{J,i}|^2}c_2\right) - \frac{|H_{s,i}|^2}{|H_{J,i}|^2}\sigma_{n,i}^2\right]^+, \quad \forall i;$
2. Allocate user signal power by
$P_{s,i}^* = \left(c_2 - \frac{|H_{J,i}|^2}{|H_{s,i}|^2}P_{J,i}^* - \sigma_{n,i}^2\right)^+, \quad \forall i.$

6.5 Numerical Results

In this section, we evaluate the impact of different strategies applied by the authorized user and the jammer on the capacity of the authorized user through numerical examples. In the following, we assume $N_c = 64$, $B = 1\,\text{MHz}$, and $P_s = P_J = 16\,\text{W}$. Both AWGN channels and frequency selective fading channels are evaluated.

6.5.1 AWGN Channels

In this subsection, we investigate AWGN channels, where the overall signal-to-noise ratio (SNR) is set to 10 dB. In light of Theorem 6.3, we assume that both the authorized user and the jammer apply uniform subchannel selection, that is, all subchannels are equally probable to be selected.

(1) **Capacity vs. Power Allocation with Fixed K_s and K_J** In this example, we evaluate the capacity of the authorized user under different transmit and jamming power allocation schemes. We set the power allocation vector as one whose elements, if sorted, would form an arithmetic sequence, and we use the *maximum power difference* among all the selected subchannels as the metric of uniformity. Hence, the maximum power difference indicates how far the power allocation is away from being uniform, and a zero difference means uniform power allocation. Figure 6.2 shows the results when both the authorized user and the jammer select half of all the available subchannels

each time, while Fig. 6.3 corresponds to the case where both of them select all the available subchannels. In the 2D view, we evaluate the capacity in two scenarios: (1) uniform jamming power allocation, while the power allocation for the authorized user is nonuniform, and (2) the case which is exactly opposite to (1). The 3D counterpart in these two figures provides spatial views on the physical meanings of the derived saddle points. Note that the saddle point is reached at one of the vertices; hence, the 3D view includes only a quarter portion of a regular saddle-point graph.

From Figs. 6.2 and 6.3, *it can be seen that*, when the number of user-activated subchannels K_s and the number of jammed subchannels K_J are both fixed: (1) if the jammer applies uniform power allocation, the authorized user maximizes its capacity when it applies uniform power allocation as well; (2) if the authorized user applies uniform power allocation, the jammer minimizes the capacity of the authorized user when it applies uniform power allocation as well; (3) the minimax capacity (the intersections in 2D view and the labeled saddle points in 3D view) serves as a lower bound when the authorized user applies uniform power allocation under all possible jamming power allocation schemes, and simultaneously it serves as an upper bound when the jammer applies uniform power allocation under all possible signal power allocation schemes. The results above match well with Theorem 6.3.

(2) **Capacity vs. Number of Selected Subchannels** In this example, we evaluate the capacity of the authorized user with different number of selected subchannels by the authorized user or the jammer. For each possible pair (K_s, K_J), both the authorized user and the jammer apply uniform power allocation. *It is observed in Fig.* 6.4 *that* the best strategy is to utilize all the N_c subchannels, either for the authorized user to maximize its capacity or for the jammer to minimize the capacity of the authorized user. This result matches well with Theorem 6.4.

6.5.2 *Frequency Selective Fading Channels*

In this subsection, we investigate frequency selective fading channels. Both the two-step water-filling algorithm for correlated fading channels and the iterative water-filling algorithm for arbitrary fading channels are evaluated.

(1) **Two-Step Water-Filling Algorithm for Correlated Fading Channels** To address the correlation between channels for the authorized user and the jammer, we introduce a correlation index, $\lambda (0 \leq \lambda \leq 1)$, which characterizes how much dependence the two channels have on each other. More specifically, in this simulation example, we generate the magnitude spectrum of channels in two steps: (1) create two random vectors, $\mathbf{x}_1 = [x_{1,1}, x_{1,2}, \ldots, x_{1,N_c}]$ and $\mathbf{x}_2 = [x_{2,1}, x_{2,2}, \ldots, x_{2,N_c}]$, in which all $x_{1,i}$ and $x_{2,i}$ are independent random variables with uniform distribution over (0,1), and (2) generate the magnitude

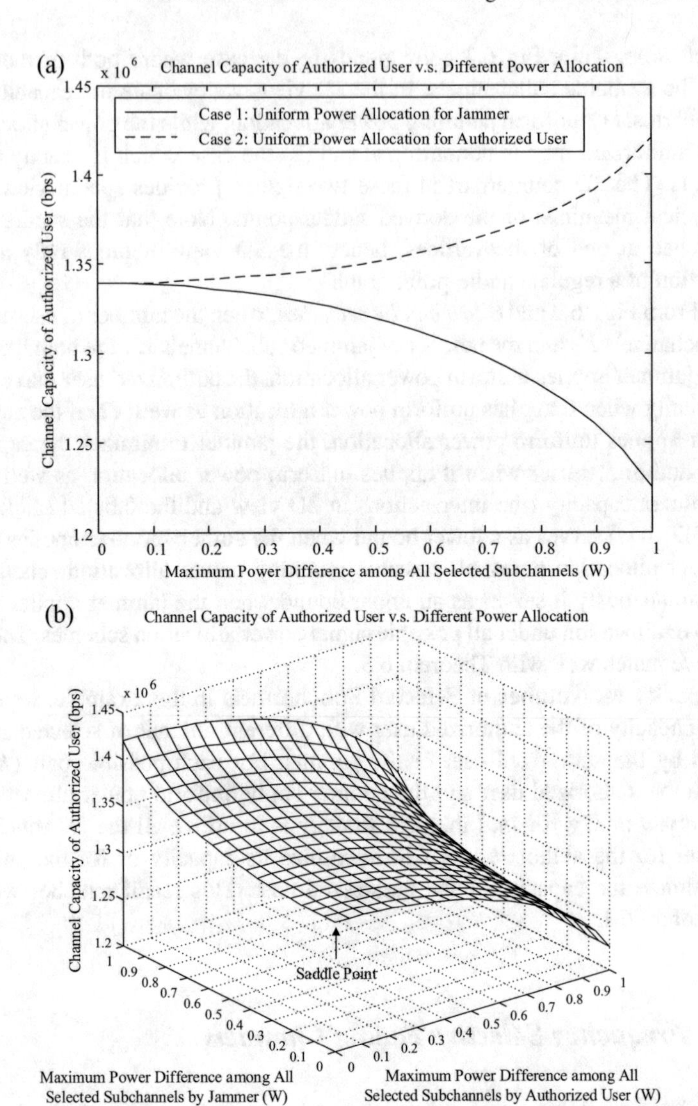

Fig. 6.2 AWGN channels: channel capacity of given bandwidth (1 MHz) vs. different power allocations. Both the authorized user and the jammer select half of all the available subchannels each time. (**a**) 2D view. (**b**) 3D view. (© [2016] IEEE. Reprinted, with permission, from Ref. [36])

spectrum of the channel for the authorized user by assigning $|H_{s,i}|^2 = x_{1,i}, \ \forall i$ and that for the jammer as $|H_{J,i}|^2 = \lambda |H_{s,i}|^2 + (1-\lambda)x_{2,i}, \ \forall i$. Particularly, $\lambda = 1$ generates equal channel magnitude spectrum for the authorized user and the jammer, while $\lambda = 0$ generates completely independent channel magnitude spectrum.

In Fig. 6.5, with the SNR being set to 10 dB, we compare the capacity of the authorized user in three cases with different power allocation strategies: (1) both

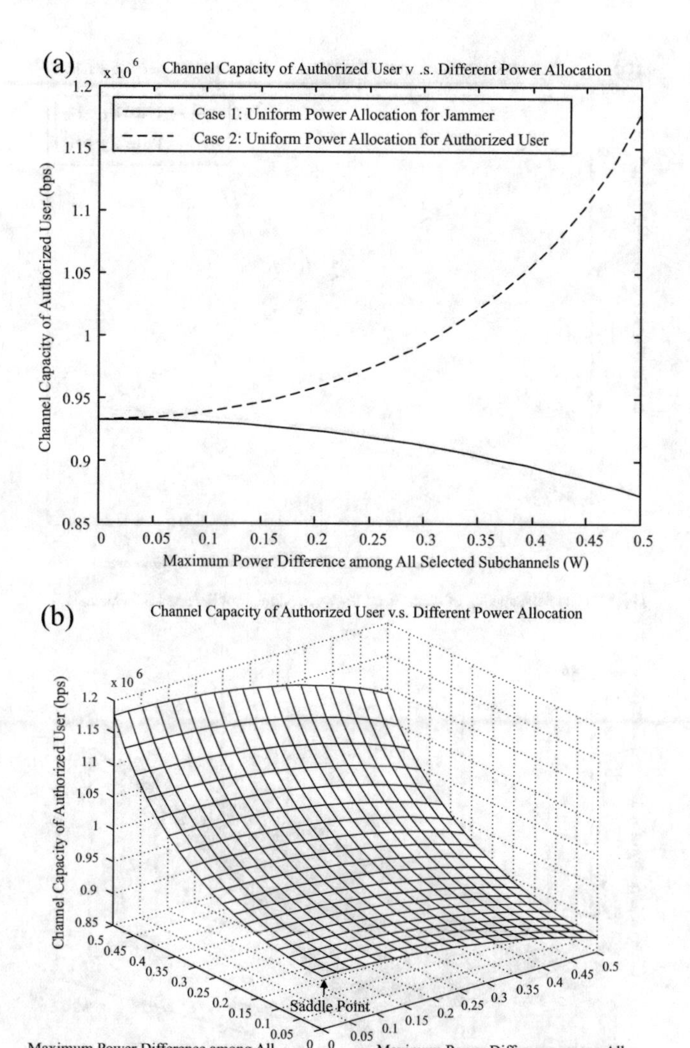

Fig. 6.3 AWGN channels: channel capacity of given bandwidth (1 MHz) vs. different power allocations. Both the authorized user and the jammer always select all the available subchannels. (**a**) 2D view. (**b**) 3D view. (© [2016] IEEE. Reprinted, with permission, from Ref. [36])

the authorized user and the jammer perform power allocation by the two-step water-filling algorithm; (2) the authorized user performs power allocation by the two-step water-filling algorithm, while the jammer performs uniform power allocation; (3) the jammer performs power allocation by the two-step water-filling algorithm, while the authorized user performs uniform power allocation.

There are *four main observations*: (1) the authorized user always has a higher capacity if he performs signal power allocation by the two-step water-filling algorithm, compared to uniform signal power allocation; (2) the capacity of the

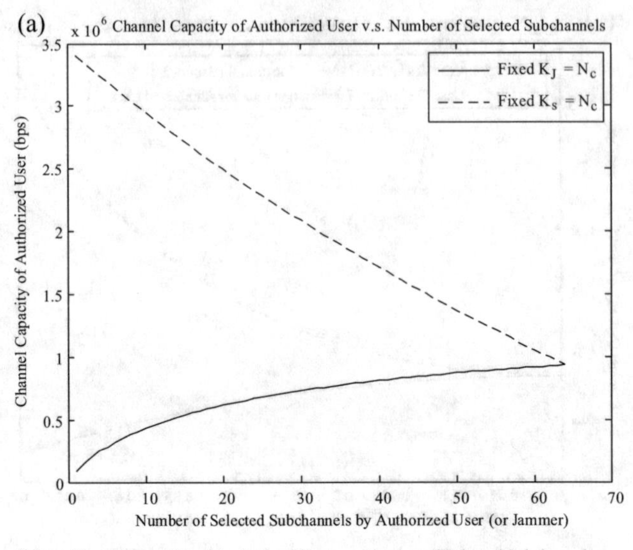

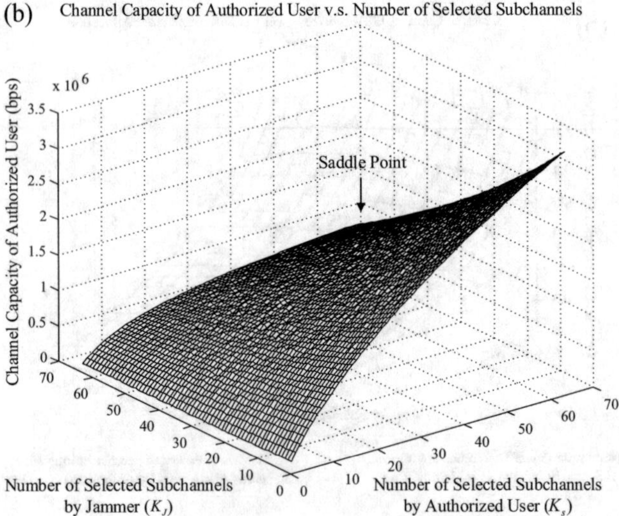

Fig. 6.4 AWGN channels: channel capacity of given bandwidth (1 MHz) vs. number of selected subchannels. (**a**) 2D view. (**b**) 3D view. (© [2016] IEEE. Reprinted, with permission, from Ref. [36])

authorized user decreases significantly if the channel of the jammer is more correlated with that of the authorized user, which implies that the jammer can enhance its jamming effect by delivering jamming power through a channel that is correlated with the authorized user's channel; (3) in a more serious case with high channel correlation, the jammer can limit the capacity of the authorized user more effectively by performing jamming power allocation by the two-step water-filling algorithm, compared to uniform jamming power allocation; (4) if

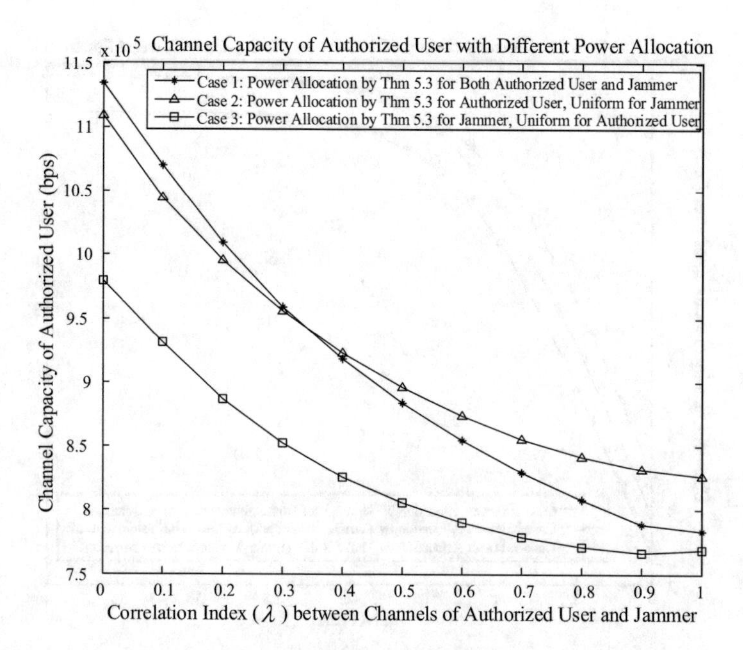

Fig. 6.5 Evaluation of the two-step water-filling algorithm under frequency selective fading channels: channel capacity of given bandwidth (1 MHz) with different power allocations vs. varying channel correlation index λ. (© [2016] IEEE. Reprinted, with permission, from Ref. [36])

the jammer is not able to achieve high channel correlation, uniform jamming power allocation is preferred instead of applying the two-step water-filling algorithm.

In Fig. 6.6, with the channel correlation index being set to $\lambda = 0.75$, we compare the capacity of the authorized user with different power allocations versus varying SNR. *It is observed that* (1) with reasonably high correlation between the user channel and the jamming channel, the power allocation strategy given by the two-step water-filling algorithm has a notable advantage over uniform power allocation, either for the authorized user to maximize its capacity or for the jammer to minimize the capacity of the authorized user, and (2) when the SNR is sufficiently high, the jamming power allocation produced by the two-step water-filling algorithm converges to uniform.

(2) **Iterative Water-Filling Algorithm for Arbitrary Fading Channels** In this simulation, the channel magnitude spectrum of the authorized user and the jammer are completely independent, which is equivalent to $\lambda = 0$ in the correlated fading channel setting. Similarly, we compare the capacity of the authorized user in three cases using the same setting as in the previous example, except that the two-step water-filling algorithm is replaced by the iterative water-filling algorithm. In Fig. 6.7, again, *it is observed that* the iterative water-filling algorithm has a notable advantage over uniform power allocation, either for the authorized user to maximize its capacity or for the jammer to minimize the capacity of the authorized user.

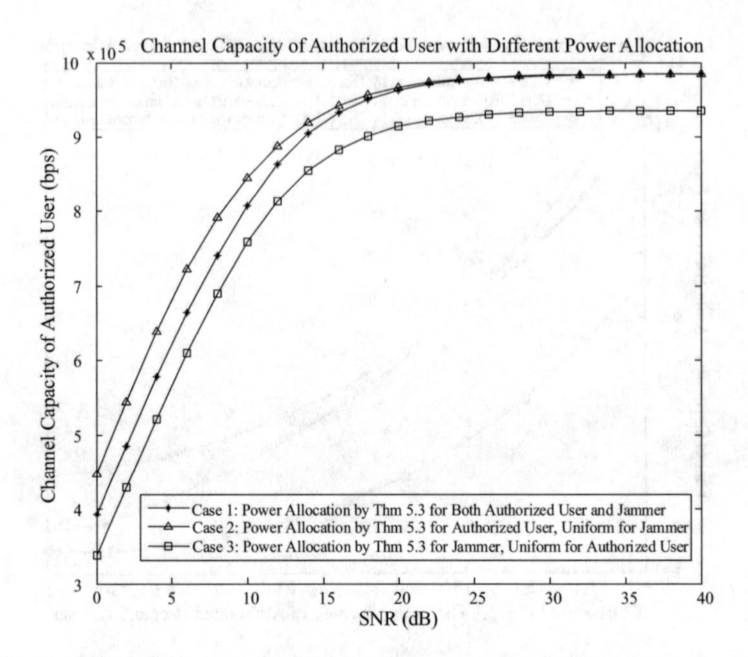

Fig. 6.6 Evaluation of the two-step water-filling algorithm under frequency selective fading channels: channel capacity of given bandwidth (1 MHz) with different power allocation vs. varying SNR. (© [2016] IEEE. Reprinted, with permission, from Ref. [36])

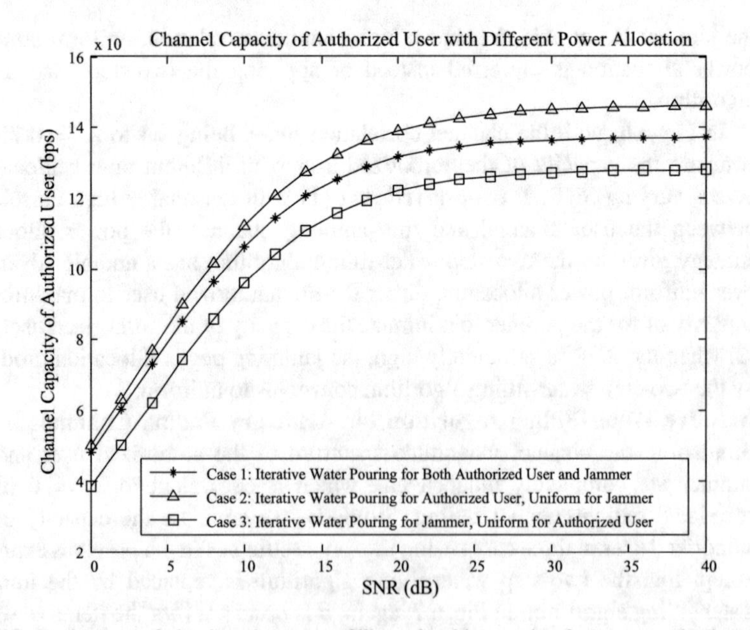

Fig. 6.7 Evaluation of the iterative water-filling algorithm under frequency selective fading channels: channel capacity of given bandwidth (1 MHz) with different power allocation vs. varying SNR. (© [2016] IEEE. Reprinted, with permission, from Ref. [36])

6.6 Conclusions

In this chapter, we considered jamming and jamming mitigation as a game between a power-limited jammer and a power-limited authorized user, who operate against each other over the same spectrum consisting of multiple bands. The strategic decision-making of the authorized user and the jammer was modeled as a two-party zero-sum game, where the payoff function is the capacity that can be achieved by the authorized user in presence of the jammer. *Under AWGN channels*, we found that either for the authorized user to maximize its capacity or for the jammer to minimize the capacity of the authorized user, the best strategy for both of them is to distribute the signal power or jamming power uniformly over all the available spectrum. *Under frequency selective fading channels*, we first characterized the dynamic relationship between the optimal signal power allocation and the optimal jamming power allocation in the minimax game and then presented an iterative water-filling algorithm to find the optimal power allocation schemes for both the authorized user and the jammer. Numerical results were provided to demonstrate the effectiveness of the optimal power allocation strategies for both AWGN and frequency selective fading channels.

References

1. R. B. Myerson, *Game Theory: Analysis of Conflict.*Cambridge, MA: Harvard University Press, 2007.
2. S. Lasaulce and H. Tembine, *Game Theory and Learning for Wireless Networks: Fundamentals and Applications*, 1st ed.Cambridge, MA: Academic Press, 2011.
3. T. Basar and Y.-W. Wu, “Solutions to a class of minimax decision problems arising in communication systems,” in *The 23rd IEEE Conference on Decision and Control*, Dec 1984, pp. 1182–1187.
4. M. Medard, “Capacity of correlated jamming channels,” in *Allerton Conference on Communications, Computing and Control*, 1997.
5. S. Farahmand, G. Giannakis, and X. Wang, “Max-min strategies for power-limited games in the presence of correlated jamming,” in *41st Annual Conference on Information Sciences and Systems*, March 2007, pp. 300–305.
6. T. Ericson, “The noncooperative binary adder channel,” *IEEE Transactions on Information Theory*, vol. 32, no. 3, pp. 365–374, 1986.
7. L. Zhang, H. Wang, and T. Li, “Anti-jamming message-driven frequency hopping-part i: System design,” *IEEE Transactions on Wireless Communications*, vol. 12, no. 1, pp. 70–79, Jan. 2013.
8. T. Song, Z. Fang, J. Ren, and T. Li, “Precoding for OFDM under disguised jamming,” in *2014 IEEE Global Communications Conference*, Dec 2014, pp. 3958–3963.
9. T. Song, K. Zhou, and T. Li, “CDMA system design and capacity analysis under disguised jamming,” *IEEE Transactions on Information Forensics and Security*, vol. PP, no. 99, pp. 1–1, 2016.
10. K. Zhou, T. Song, J. Ren, and T. Li, “Robust CDMA receiver design under disguised jamming,” in *2016 IEEE International Conference on Acoustics, Speech and Signal Processing (ICASSP)*, March 2016, pp. 2179–2183.

11. J. Chiang and Y.-C. Hu, "Cross-layer jamming detection and mitigation in wireless broadcast networks," *IEEE/ACM Transactions on Networking*, vol. 19, no. 1, pp. 286–298, Feb 2011.
12. R. Di Pietro and G. Oligeri, "Jamming mitigation in cognitive radio networks," *IEEE Network*, vol. 27, no. 3, pp. 10–15, May 2013.
13. T. Basar, "The Gaussian test channel with an intelligent jammer," *IEEE Transactions on Information Theory*, vol. 29, no. 1, pp. 152–157, Jan 1983.
14. S. Diggavi and T. Cover, "The worst additive noise under a covariance constraint," *IEEE Transactions on Information Theory*, vol. 47, no. 7, pp. 3072–3081, Nov 2001.
15. W. Stark and R. McEliece, "On the capacity of channels with block memory," *IEEE Transactions on Information Theory*, vol. 34, no. 2, pp. 322–324, Mar 1988.
16. J. Borden, D. Mason, and R. McEliece, "Some information theoretic saddlepoints," *SIAM Journal on Control and Optimization*, vol. 23, no. 1, pp. 129–143, 1985.
17. R. Mallik, R. Scholtz, and G. Papavassilopoulos, "Analysis of an on-off jamming situation as a dynamic game," *IEEE Transactions on Communications*, vol. 48, no. 8, pp. 1360–1373, Aug 2000.
18. G. Scutari, D. Palomar, and S. Barbarossa, "Optimal linear precoding strategies for wideband noncooperative systems based on game theory-part i: Nash equilibria," *IEEE Transactions on Signal Processing*, vol. 56, no. 3, pp. 1230–1249, March 2008.
19. Z. Han, Z. Ji, and K. Liu, "Fair multiuser channel allocation for OFDMA networks using Nash bargaining solutions and coalitions," *IEEE Transactions on Communications*, vol. 53, no. 8, pp. 1366–1376, Aug 2005.
20. A. Garnaev, Y. Hayel, and E. Altman, "A bayesian jamming game in an OFDM wireless network," in *10th International Symposium on Modeling and Optimization in Mobile, Ad Hoc and Wireless Networks*, May 2012, pp. 41–48.
21. R. Gohary, Y. Huang, Z.-Q. Luo, and J.-S. Pang, "A generalized iterative water-filling algorithm for distributed power control in the presence of a jammer," *IEEE Transactions on Signal Processing*, vol. 57, no. 7, pp. 2660–2674, July 2009.
22. Z. Han, N. Marina, M. Debbah, and A. Hjørungnes, "Physical layer security game: Interaction between source, eavesdropper, and friendly jammer," *EURASIP Journal on Wireless Communications and Networking*, vol. 2009, no. 1, pp. 1–10, 2010.
23. J. Neel, R. Buehrer, J. Reed, and R. P. Gilles, "Game theoretic analysis of a network of cognitive radios," in *45th Midwest Symposium on Circuits and Systems*, vol. 3, Aug 2002, pp. III–409–III–412 vol.3.
24. Z. Ji and K. Liu, "Dynamic spectrum sharing: A game theoretical overview," *IEEE Communications Magazine*, vol. 45, no. 5, pp. 88–94, May 2007.
25. R. El-Bardan, S. Brahma, and P. Varshney, "A game theoretic power control framework for spectrum sharing in competitive environments," in *Asilomar Conference on Signals, Systems and Computers*, Nov 2013, pp. 1493–1497.
26. V. Nadendla, H. Chen, and P. Varshney, "Minimax games for cooperative spectrum sensing in a centralized cognitive radio network in the presence of interferers," in *MILITARY COMMUNICATIONS CONFERENCE*, Nov 2011, pp. 1256–1260.
27. R. Menon, A. MacKenzie, R. Buehrer, and J. Reed, "A game-theoretic framework for interference avoidance in ad hoc networks," in *IEEE Global Telecommunications Conference*, Nov 2006, pp. 1–6.
28. S. Wei, R. Kannan, V. Chakravarthy, and M. Rangaswamy, "Csi usage over parallel fading channels under jamming attacks: A game theory study," *IEEE Transactions on Communications*, vol. 60, no. 4, pp. 1167–1175, April 2012.
29. A. Garnaev, W. Trappe, and A. Petropulu, "Equilibrium strategies for an ofdm network that might be under a jamming attack," in *2017 51st Annual Conference on Information Sciences and Systems (CISS)*, March 2017, pp. 1–6.
30. P. D. Straffin, *Game Theory and Strategy*, 1st ed. Washinton, DC: The Mathematical Association of America, 1993.
31. R. Ahlswede, "Elimination of correlation in random codes for arbitrarily varying channels," *Z. Wahrscheinlichkeitstheorie und Verwandte Gebiete*, vol. 44, no. 2, pp. 159–175, 1978.

32. A. Lapidoth and P. Narayan, "Reliable communication under channel uncertainty," *IEEE Transactions on Information Theory*, vol. 44, no. 6, pp. 2148–2177, 1998.
33. L. Zhang and T. Li, "Anti-jamming message-driven frequency hopping-part ii: Capacity analysis under disguised jamming," *IEEE Transactions on Wireless Communications*, vol. 12, no. 1, pp. 80–88, Jan. 2013.
34. S. Boyd and L. Vandenberghe, *Convex Optimization*.Cambridge University Press, 2004.
35. Q. Qi, A. Minturn, and Y. Yang, "An efficient water-filling algorithm for power allocation in OFDM-based cognitive radio systems," in *2012 International Conference on Systems and Informatics (ICSAI)*, May 2012, pp. 2069–2073.
36. T. Song, W. E. Stark, T. Li, and J. K. Tugnait, "Optimal multiband transmission under hostile jamming," *IEEE Transactions on Communications*, vol. 64, no. 9, pp. 4013–4027, Sept 2016.
37. W. Yu, W. Rhee, S. Boyd, and J. M. Cioffi, "Iterative water-filling for Gaussian vector multiple-access channels," *IEEE Transactions on Information Theory*, vol. 50, no. 1, pp. 145–152, Jan 2004.
38. W. Yu and J. M. Cioffi, "On constant power water-filling," in *IEEE International Conference on Communications*, vol. 6, 2001, pp. 1665–1669.

Chapter 7
Conclusions and Future Directions

7.1 What We Learned About Jamming and Anti-jamming

In this section, we summarize what we have learned from our study on jamming and anti-jamming. As the old saying goes, "precise knowledge of self and precise knowledge of the threat leads to victory," a clear understanding on jamming is the prerequisite for effective anti-jamming system design and performance evaluation.

7.1.1 *On Jamming*

For jamming, we learned that:

- *Disguised jamming, which only has the same power level as the authorized signal, can be more harmful than partial band strong jamming.* For a long time, it was believed that only strong jamming is really harmful. It turns out that this is only partially true. If the authorized signal coding space and pulse shaping filter are observed by the jammer, the jammer can then mimic the characteristics of the authorized signal and launch disguised jamming which is of the same power level as that of the authorized signal, such that the receiver cannot distinguish the jamming from the authorized signal. The symmetricity between the jamming and the authorized signal leads to complete communication failure. To make things worse, disguised jamming attack happens at the symbol level and is immune to bit-level channel coding. That is, reducing the code rate at the bit level will not improve the system performance under disguised jamming.

  Assuming the jammer and the authorized user have the same power constraint, to achieve strong jamming, the jammer has to concentrate its power on certain frequency bands. On the other hand, with disguised jamming, the jammer can

T. Li et al., *Wireless Communications Under Hostile Jamming: Security and Efficiency*, https://doi.org/10.1007/978-981-13-0821-5_7

easily launch full-band jamming, which is highly devastating to communication systems.

- *The ultimate game between the user and the jammer leads to uniform power distribution over the whole band from both sides.* Jamming can be highly cognitive. When a power-limited authorized user and a power-limited jammer are operating swiftly and independently over the same spectrum consisting of multiple bands, it would be impossible to track the opponent's transmission pattern. In this case, the strategic decision-making of the authorized user and the jammer can be modeled as a two-party zero-sum game, where the payoff function is the capacity that can be achieved by the authorized user in presence of the jammer. We investigated the game under additive white Gaussian noise (AWGN) channels and explore the possibility for the authorized user or the jammer to randomly utilize part (or all) of the available spectrum and/or apply nonuniform power allocation. It was found that under AWGN channels, either for the authorized user to maximize its capacity or for the jammer to minimize the capacity of the authorized user, the best strategy is to distribute the transmission power or jamming power uniformly over all the available spectrum. The minimax capacity can be calculated based on the channel bandwidth and the signal-to-jamming and noise ratio, and it matches with the Shannon channel capacity formula.

7.1.2 On Anti-jamming System Design

For anti-jamming, we learned that:

- *Existing communication systems do not possess sufficient security features and are generally very fragile under disguised jamming.* Most communication systems, such as EDGE and OFDM, are designed for efficient and accurate information transmission from the source to destination and do not have inherent security features. When attacked by disguised jamming, the deterministic code capacity of these systems will be reduced to zero. Existing work on anti-jamming system design or jamming mitigation is mainly based on spread spectrum techniques, including CDMA and frequency hopping (FH) [1–6]. Both CDMA and FH systems possess anti-jamming features by exploiting frequency diversity over large spectrum [7]. However, while these systems work reasonably well for voice-centric communication, their security feature and information capacity are far from adequate and acceptable for today's high-speed multimedia wireless services. Their security is only assured by the PN sequence generators; their spectral efficiency is largely limited by self-jamming and repeated coding.

  Unlike eavesdropping, which is a passive attack, jamming is an active attack which cannot be mitigated based solely on higher-layer security measures but has to be confronted directly from the physical layer. In short, we have to develop new transceiver design techniques so that the wireless systems have inherent, built-in security features.

- *Secure yet efficient anti-jamming wireless communications can be achieved by integrating advanced cryptographic techniques into physical layer transceiver design.* Over the last 15 years, we came up with secure and efficient anti-jamming designs for all the major communication systems:
  - We enhanced the security of CDMA systems through secure scrambling; in addition to its robustness to narrow-band jamming, CDMA with secure scrambling is proved to be secure and efficient under disguised jamming. At the same time, it has comparable complexity as the traditional CDMA system.
  - We developed innovative frequency hopping techniques, known as message-driven frequency hopping (MDFH) and collision-free frequency hopping (CFFH). In MDFH, part of the message stream acts as the PN sequence and is transmitted through hopping frequency control. As a result, system efficiency is increased significantly since additional information transmission is achieved at no extra cost on either bandwidth or power. Quantitative performance analysis on the presented schemes demonstrates that transmission through hopping frequency control essentially introduces another dimension to the signal space and the corresponding coding gain can increase the system efficiency by *multiple* times.

    On the other hand, CFFH was developed to resolve the self-collision problem in conventional frequency hopping. Based on an innovative secure subcarrier assignment scheme, CFFH was designed to ensure that (i) each user hops to a new set of subcarriers in a pseudorandom manner at the beginning of each hopping period, (ii) different users always transmit on non-overlapping sets of subcarriers, and (iii) malicious users cannot predict or repeat the hopping pattern of the authorized users and hence cannot launch follower jamming attacks. CFFH can be directly applied to any multiband multiaccess systems to achieve secure and dynamic spectral access control. When applied to OFDMA system, CFFH is proved to be particularly effective in mitigating partial band jamming and follower jamming. At the same time, CFFH-OFDM has the same spectral efficiency as the original OFDM system.
  - We developed an innovative secure precoding scheme which can achieve constellation randomization, break the symmetricity between the authorized signal and jamming, and hence achieve secure and efficient transmission under disguised jamming. We applied it to the OFDM system and came up with the securely precoded OFDM, known as SP-OFDM. It was proved that while achieving strong resistance against disguised jamming, SP-OFDM has the same high spectral efficiency as the traditional OFDM system. Moreover, we showed that due to the inherent secure randomness in SP-OFDM, disguised jamming is no longer the worst jamming for SP-OFDM.

7.2 Discussions on Future Directions

In this section, we discuss some open research topics concerning the physical layer and network layer design of 5G networks under hostile environments.

7.2.1 Jamming and Anti-jamming in 5G IoT Systems

The Internet of things (IoT), which networks versatile devices for information exchange, remote sensing, monitoring, and control, is finding promising applications in nearly every field. As can be seen, the key enabler for the IoT would be seamless, secure, and reliable wireless communications. The enormous spectrum required by the IoT seems to be not available until the reveal of the 5G-millimeter wave (mmWave) band (28–300 GHz). Compared to existing lower band systems (such as 3G, 4G), mmWave band signals generally require line of sight (LOS) path and suffer from severe fading effects, leading to much smaller coverage area. That is, 5G networks will have very high node density. This implies that when hostile jamming is present, it is likely to have direct, destructive impact over a large number of nodes within its power zone.

However, “when a door closes, a window opens”; the good news is that while 5G has high node density, it also has a huge frequency range which possesses enormous frequency diversity and makes it impossible for the jammer to launch a full-band jamming. This implies that we can defend our networks against jamming through dynamic spectrum access control for nodes of different layers over a very large frequency range using collision-free frequency hopping, we can combat disguised jamming using secure constellation randomization, and we can also improve user coverage and network capacity through overlapping user grouping in 5G systems [8–10]. With that being said, anti-jamming system design for 5G IoT networks is surely an important and challenging yet tractable task.

7.2.2 5G Wireless Network Design and Performance Evaluation Under Hostile Environments

Effective network operations face serious threats from malicious attacks, especially active attacks like jamming, Byzantine, and various routing attacks [11–15]. In this section, we will discuss network failure detection, impact of the failure on network performance, and how to ensure quality transmission in hostile environments through diversity enhancement and dynamic protocol development.

7.2.2.1 Network Failure Detection and Network Performance Evaluation Under Malicious Attacks

Network failure detection Network failure detection in multihop networks is a challenging task. When an erroneous transmission is detected, it could be a compromised node which sends fake information or launches routing attacks by dropping packets or modifying their contents as they are being forwarded [16], or it can be a link failure caused by malicious jamming. Consider a multihop network where there are multiple relays at each hop level to assist the transmission. It can be shown that if there is at least one normal relay in each hop, then we can detect all the suspicious nodes between the source and the destination. When a "suspicious" node is identified, it will be excluded from the active routing paths. We can then further analyze that the erroneous transmission is caused by the node failure or link failure.

Network performance evaluation and monitoring under malicious attacks The performance of a network is determined by the performance of each node and the connections between them. As different nodes play different roles in the network, some nodes are more "important" than others. For this reason, we introduce the concept of node significance. The significance of a node is determined by the intensity of the overall traffic associated to it, including self-generated traffic, relay traffic, and received/stored traffic. Given a particular node k, which can either be a basic node or a relay or base station, for accurate characterization of its significance, all the nodes connected to it need to be considered. Let $T_{i,j}^{(k)}$ denote the throughput from source i to destination j that passes through node k. When $i = k$, $T_{i,j}^{(k)}$ is the self-generated traffic of node k that is received at node j; when $j = k$, $T_{i,j}^{(k)}$ is the received/stored traffic at node k from node i. Let $Pr(i \to k \to j)$ denote the probability that path $i \to k \to j$ has been taken; then the averaged total traffic associated with node k can be represented as

$$T^{(k)} = \sum_{i,j} T_{i,j}^{(k)} Pr(i \to k \to j). \tag{7.1}$$

Obviously, the larger the $T^{(k)}$, the more significant the node. Under malicious attacks, the total traffic lost can be estimated as $\sum_k T^{(k)} Pr(\text{node k is "lost"})$. For network performance evaluation, we can check the total traffic flow and stability at the most significant nodes (such as the relay nodes and the sinks) periodically; we can also monitor the delays in packet delivery from the basic node to the sink by inserting time stamps in the packets. We also need to investigate how exceptional changes in throughput, traffic flow pattern, and delay reflect the real-time status of the network performance.

7.2.2.2 Improving Network Reliability Through Topology Design and Dynamic Routing Protocol Development

Network reliability can be improved through hop-number control and diversity enhancement. One needs to explore how to achieve that through topology design, node deployment, and routing protocol development, with considerations on the trade-off between efficiency and security. The basic idea is to optimize the network throughput and delay performance, but at the same time, to ensure sufficient diversity such that the network will not get paralyzed as long as the node/link damages are within a certain limit.

Topology diversity and network architecture design Note that topology diversity determines the minimum effort required to break down a network. Here one needs to answer the following questions: (i) Given a fixed number of relay nodes, how to deploy them so as to achieve the maximum number of path or maximum routing diversity under certain delay constraints? (ii) If the percentage of malicious node or failed link is given, how to design the network so that the probability of successful transmission is above a certain threshold?

Given that 5G is a high density heterogeneous wireless system, the topology design of 5G networks will reflect the convergence of centralized and ad hoc networks. That is, 5G network needs to have a well-organized infrastructure to ensure the reliability (including both transmission accuracy and security), capacity, energy efficiency, as well as time efficiency. At the same time, the network should provide sufficient flexibility by allowing authorized ad hoc communications among the nodes or devices. A unified framework for wireless network characterization can be found in [17]. It includes all existing network models as special cases and can serve as a benchmark for 5G network design and performance analysis.

Dynamic routing Dynamic routing is one of the most effective methods to mitigate various routing attacks, including both static and dynamic attacks. Under static attacks, secure routing path selection can be performed by exploiting malicious node detection. When the attacks are time varying, the situation becomes very complex.

One interesting task is to develop a routing protocol that can maximize the message delivery ratio under dynamic attacks, subject to predefined energy and delay constraints. We can combine random walking and deterministic routing for packet transmission. The dynamic adoption of the two strategies can be determined based on the security policy and knowledge of the network diversity. We need to minimize the probability for different packets transmitted between the same source and destination pair to be routed through the same routing path, such that attackers cannot locate the packet; at the same time, the authorized user who has knowledge of the selection strategy and network diversity can easily identify and get the packets. We need to address the following questions: (i) How to achieve secure networking by introducing real-time secure randomness into the network through dynamic node status control, physical layer modulation control, dynamic scheduling, and routing (i.e., dynamic networking)? (ii) How to characterize or measure the security level of a routing protocol? (iii) How to quantify the design trade-off between security and efficiency under adversarial attacks?

7.3 Concluding Remarks

The battle between authorized communications and malicious attacks will never end. For the upcoming 5G system, the most significant features are enormous spectrum, large scale, and high node density. The vast spectrum will naturally provide rich spectral diversity, which can be exploited to enhance jamming resistance. The large size and high density can also help increase the network reliability. As an illustration, in our previous research [18] on distributed detection under Byzantine attacks, we proved that the detection error probability diminishes exponentially as the network size increases, even if the percentage of malicious nodes remains fixed. This implies that network performance under malicious attacks relies heavily on the network density or size, which happens to be the strength of 5G systems.

Another notable attribute of 5G is that it operates in the mmWave bands, which means the antenna size is very small. This makes it possible to equip a large number (several tens or even hundreds) of antennas at both the transmitter and receiver sides, an efficient transmission technique known as massive multi-input multi-output (massive MIMO). In addition to frequency and time diversity, potentially, the spatial diversity introduced by massive MIMO and node location randomness can be exploited for secure communications under malicious environments.

Along with the rapid advances in cloud computing, distributed storage, block chain, artificial intelligence, and other related technologies, we are optimistic that the human society will be able to establish a powerful and highly intelligent cyberphysical system enabled by secure and reliable wireless communication networks.

References

1. R.V. Alred. Naval radar anti-jamming technique. *Journal of the Institution of Electrical Engineers - Part III A: Radiolocation*, 93(10):1593–1601, 1946.
2. M.G. Amin. Interference mitigation in spread spectrum communication systems using time-frequency distributions. *IEEE Transactions on Signal Processing*, 45(1):90–101, 1997.
3. Wei Sun and M.G. Amin. A self-coherence anti-jamming gps receiver. *IEEE Transactions on Signal Processing*, 53(10):3910–3915, 2005.
4. I. Bergel, E. Fishler, and H. Messer. Narrowband interference mitigation in impulse radio. *IEEE Transactions on Communications*, 53(8):1278–1282, 2005.
5. Yongle Wu, Beibei Wang, K.J.R. Liu, and T.C. Clancy. Anti-jamming games in multi-channel cognitive radio networks. *IEEE Journal on Selected Areas in Communications*, 30(1):4–15, 2012.
6. C. Popper, M. Strasser, and S. Capkun. Anti-jamming broadcast communication using uncoordinated spread spectrum techniques. *IEEE Journal on Selected Areas in Communications*, 28(5):703–715, 2010.
7. R.C. Dixon. *Spread Spectrum Systems with Commercial Applications*. John Wiley & Son, Inc, third edition, 1994.
8. Run Tian, Yuan Liang, Xuezhi Tan, and Tongtong Li. Overlapping user grouping in iot oriented massive mimo systems. *IEEE Access*, 5:14177–14186, 2017.

9. H. Q. Ngo, E. G. Larsson, and T. L. Marzetta. Aspects of favorable propagation in massive MIMO. In *European Signal Processing Conference*, pages 76–80, September 2014.
10. Ulrike Von Luxburg. A tutorial on spectral clustering. *Statistics and computing*, 17(4):395–416, 2007.
11. A.S. Rawat, P. Anand, Hao Chen, and P.K. Varshney. Collaborative spectrum sensing in the presence of byzantine attacks in cognitive radio networks. *IEEE Transactions on Signal Processing*, 59(2):774–786, Feb. 2011.
12. Yanchao Zhang, Wei Liu, Wenjing Lou, and Yuguang Fang. Location-based compromise-tolerant security mechanisms for wireless sensor networks. *IEEE Journal on Selected Areas in Communications*, 24(2):247–260, Feb.
13. Mai Abdelhakim, Leonard Lightfoot, Jian Ren, and T. Li. Distributed detection in mobile access wireless sensor networks under byzantine attacks. *IEEE Transactions on Parallel and Distributed Systems*, 99, 2013.
14. Lei Zhang and Tongtong Li. Anti-jamming message-driven frequency hopping; part ii: Capacity analysis under disguised jamming. *IEEE Transactions on Wireless Communications*, 12(1):80–88, 2013.
15. R. Niu and P.K. Varshney. Performance analysis of distributed detection in a random sensor field. *IEEE Transactions on Signal Processing*, 56(1):339–349, Jan. 2008.
16. C. Karlof and D. Wagner. Secure routing in wireless sensor networks: attacks and countermeasures. *IEEE International Workshop on Sensor Network Protocols and Applications*, pages 113–127, May 2003.
17. Tongtong Li, M. Abdelhakim, and Jian Ren. N-hop networks: a general framework for wireless systems. *IEEE Wireless Communications*, 21(2):98–105, April 2014.
18. M. Abdelhakim, L.E. Lightfoot, Jian Ren, and Tongtong Li. Distributed detection in mobile access wireless sensor networks under Byzantine attacks. *IEEE Transactions on Parallel and Distributed Systems*, 25(4):950–959, April 2014.

Appendix A
Proof of Lemma 3.3

Note that $\mathcal{X}$ can be partitioned into six subsets with respect to $\mathbf{x}_0 = \boldsymbol{\alpha} s$. Define $\mathcal{B}_1 \triangleq \{\boldsymbol{\alpha}\tilde{s}|\tilde{s} \in \Omega\}$, $\mathcal{B}_2 \triangleq \{\boldsymbol{\beta}\tilde{s}|\tilde{s} \in \Omega\}$ and $\mathcal{B}_3 \triangleq \{\boldsymbol{\alpha}_0\tilde{s}|\tilde{s} \in \Omega, \boldsymbol{\alpha}_0 \neq \boldsymbol{\alpha}, \boldsymbol{\beta}\}$. We have $\mathcal{X} = \cup_{i=1}^{3}\mathcal{B}_i$. It follows from the definition of subset $\mathcal{X}_i$ in (3.103) that $\mathcal{B}_1 = \mathcal{X}_1 \cup \mathcal{X}_2$, $\mathcal{B}_2 = \mathcal{X}_3 \cup \mathcal{X}_4$, $\mathcal{B}_3 = \mathcal{X}_5 \cup \mathcal{X}_6$, and $\mathcal{X} = \cup_{i=1}^{6}\mathcal{X}_i$.

(i) We consider the cases where $\mathbf{y} \in \mathcal{X}_i$, $i = 1, 3$. When $\mathbf{y} \in \mathcal{X}_1$ ($\mathbf{y} = -\mathbf{x}_0$), the jamming cancels the true signal, and the received signal contains only noise, resulting in $W(\hat{\boldsymbol{\alpha}}_0|\mathbf{x}_0, -\mathbf{x}_0) = \frac{1}{N_c}, \forall\hat{\boldsymbol{\alpha}}_0 \in \mathcal{A}$. When $\mathbf{y} \in \mathcal{X}_3$ ($\mathbf{y} = \boldsymbol{\beta} s$), the jamming has the same ID symbol as the true signal, and the receiver cannot distinguish between the two, resulting in $W(\boldsymbol{\alpha}|\mathbf{x}_0, \boldsymbol{\beta} s) = W(\boldsymbol{\beta}|\mathbf{x}_0, \boldsymbol{\beta} s)$. Hence, $W(\boldsymbol{\alpha}|\mathbf{x}_0, \mathbf{y}) = W(\boldsymbol{\beta}|\mathbf{x}_0, \mathbf{y})$ holds in both cases.

(ii) When $\mathbf{y} \in \mathcal{X}_2$, we have $\mathbf{y} = \boldsymbol{\alpha} s_0$ where $s_0 \in \Omega$, $s_0 \neq -s$. Assuming $\boldsymbol{\alpha} = v(k)$,

$$\begin{aligned} W(\boldsymbol{\alpha}|\mathbf{x}_0, \mathbf{y}) &= Pr\{Z_k < Z_i, \forall i \in \mathcal{I}_c, i \neq k|\mathbf{x}_0, \mathbf{y}\} \\ &\geq 1 - \sum_{i \neq k} Pr\{Z_k \geq Z_i|\mathbf{x}_0, \mathbf{y}\} \\ &= 1 - (N_c - 1)Pr\{Z_k \geq Z_{i_0}|\mathbf{x}_0, \mathbf{y}\}, \end{aligned} \tag{A.1}$$

for any fixed $i_0 \neq k$. Since $s_0 \neq -s$, it follows from the results in [1] that

$$\begin{aligned} Pr\{Z_k \geq Z_{i_0}|\mathbf{x}_0, \mathbf{y}\} &= Q_1(\sqrt{C}, \sqrt{D}) - \frac{\|s + s_0\|^2 + \sigma_n^2}{\|s + s_0\|^2 + 2\sigma_n^2} e^{-\frac{C+D}{2}} I_0(\sqrt{CD}) \\ &< Q_1(\sqrt{C}, \sqrt{D}), \end{aligned} \tag{A.2}$$

where $C = \frac{2P_s}{\|s+s_0\|^2+2\sigma_n^2}$ and $D = \frac{2P_s(\|s+s_0\|^2+\sigma_n^2)}{\sigma_n^2(\|s+s_0\|^2+2\sigma_n^2)}$. Since $D > C$, $Pr\{Z_k \geq Z_{i_0}|\mathbf{x}_0, \mathbf{y}\} \leq e^{-\frac{(D-C)^2}{2}}$ [2]. It then follows from (A.1) and

T. Li et al., *Wireless Communications Under Hostile Jamming: Security and Efficiency*, https://doi.org/10.1007/978-981-13-0821-5

$\|s + s_0\| \geq d_{\min}$ that $W(\boldsymbol{\alpha}|\mathbf{x}_0, \mathbf{y}) > 1 - (N_c - 1)\exp[-2(\frac{\gamma d_{\min}^2}{d_{\min}^2+2\sigma_n^2})^2]$, and $W(\boldsymbol{\beta}|\mathbf{x}_0, \mathbf{y}) = \frac{1}{N_c-1}[1 - W(\boldsymbol{\alpha}|\mathbf{x}_0, \mathbf{y})] < \exp[-2\left(\frac{\gamma d_{\min}^2}{d_{\min}^2+2\sigma_n^2}\right)^2]$. Hence, we have $W(\boldsymbol{\alpha}|\mathbf{x}_0, \mathbf{y}) - W(\boldsymbol{\beta}|\mathbf{x}_0, \mathbf{y}) > 1 - N_c\exp[-2\left(\frac{\gamma d_{\min}^2}{d_{\min}^2+2\sigma_n^2}\right)^2]$, which implies that under the conditions

$$\gamma > \sqrt{\frac{1}{2}\ln N_c} \text{ and } \frac{d_{\min}^2}{\sigma_n^2} > \frac{2\sqrt{\ln N_c}}{\sqrt{2}\gamma - \sqrt{\ln N_c}}, \tag{A.3}$$

$W(\boldsymbol{\alpha}|\mathbf{x}_0, \mathbf{y}) - W(\boldsymbol{\beta}|\mathbf{x}_0, \mathbf{y}) > 0$.

(iii) When $\mathbf{y} \in \mathcal{X}_4$, we have $\mathbf{y} = \boldsymbol{\beta}s_0$ where $s_0 \in \Omega, s_0 \neq s$. Note that $W(\boldsymbol{\alpha}|\mathbf{y}, \mathbf{x}_0) = W(\boldsymbol{\beta}|\mathbf{x}_0, \mathbf{y})$. Since $\|s_0 - s\| \geq d_{\min}$, it follows from Theorem 3.2 that $W(\boldsymbol{\alpha}|\mathbf{x}_0, \mathbf{y}) - W(\boldsymbol{\beta}|\mathbf{x}_0, \mathbf{y}) \geq 1 - e^{-\frac{d_{\min}^2}{2\sigma_n^2}} - 2\epsilon$. Therefore, under the conditions,

$$\epsilon < \frac{1}{2} \text{ and } \frac{d_{\min}^2}{\sigma_n^2} > 2\ln\frac{1}{1-2\epsilon}, \tag{A.4}$$

$W(\boldsymbol{\alpha}|\mathbf{x}_0, \mathbf{y}) - W(\boldsymbol{\beta}|\mathbf{x}_0, \mathbf{y}) > 0$.

(iv) When $\mathbf{y} \in \mathcal{X}_5$, we have $\mathbf{y} = \boldsymbol{\alpha}_0 s$ where $\boldsymbol{\alpha}_0 \in \mathcal{A}, \boldsymbol{\alpha}_0 \neq \boldsymbol{\alpha}, \boldsymbol{\beta}$. It follows from Proposition 3.5 that $W(\boldsymbol{\alpha}|\mathbf{x}_0, \mathbf{y}) \geq \frac{1}{2} - \epsilon$. Note that $W(\boldsymbol{\alpha}|\mathbf{x}_0, \mathbf{y}) = W(\boldsymbol{\alpha}_0|\mathbf{x}_0, \mathbf{y})$, we have $W(\boldsymbol{\beta}|\mathbf{x}_0, \mathbf{y}) = \frac{1}{N_c-2}[1 - 2W(\boldsymbol{\alpha}|\mathbf{x}_0, \mathbf{y})] \leq \frac{2\epsilon}{N_c-2}$. Hence, we have $W(\boldsymbol{\alpha}|\mathbf{x}_0, \mathbf{y}) - W(\boldsymbol{\beta}|\mathbf{x}_0, \mathbf{y}) \geq \frac{1}{2} - \frac{N_c\epsilon}{N_c-2}$. Under the condition

$$\epsilon < \frac{N_c - 2}{2N_c}, \tag{A.5}$$

$W(\boldsymbol{\alpha}|\mathbf{x}_0, \mathbf{y}) - W(\boldsymbol{\beta}|\mathbf{x}_0, \mathbf{y}) > 0$.

(v) When $\mathbf{y} \in \mathcal{X}_6$, we have $\mathbf{y} = \boldsymbol{\alpha}_0 s_0$ where $\boldsymbol{\alpha}_0 \in \mathcal{A}, \boldsymbol{\alpha}_0 \neq \boldsymbol{\alpha}, \boldsymbol{\beta}$ and $s_0 \in \Omega, s_0 \neq s$. It follows from Proposition 3.5 that $W(\boldsymbol{\alpha}|\mathbf{x}_0, \mathbf{y}) \geq 1 - \frac{1}{2}e^{-\frac{\|s_0-s\|^2}{2\sigma_n^2}} - \epsilon$. Assuming $\boldsymbol{\alpha} = v(k)$ and $\boldsymbol{\alpha}_0 = v(k_0)$, it follows from Lemma 3.1 that $W(\boldsymbol{\alpha}_0|\mathbf{x}_0, \mathbf{y}) \geq Pr\{Z_{k_0} < Z_k|\mathbf{x}_0, \mathbf{y}\} - (N_c - 2)Pr\{Z_{k_0} \geq Z_{i_0}|\mathbf{x}_0, \mathbf{y}\} = \frac{1}{2}e^{-\frac{\|s_0-s\|^2}{2\sigma_n^2}} - \epsilon$. Then we have $W(\boldsymbol{\beta}|\mathbf{x}_0, \mathbf{y}) = \frac{1}{N_c-2}[1 - W(\boldsymbol{\alpha}|\mathbf{x}_0, \mathbf{y}) - W(\boldsymbol{\alpha}_0|\mathbf{x}_0, \mathbf{y})] \leq \frac{2\epsilon}{N_c-2}$. Hence, we have $W(\boldsymbol{\alpha}|\mathbf{x}_0, \mathbf{y}) - W(\boldsymbol{\beta}|\mathbf{x}_0, \mathbf{y}) \geq 1 - \frac{1}{2}e^{-\frac{d_{\min}^2}{2\sigma_n^2}} - \frac{N_c\epsilon}{N_c-2}$, which implies that under the conditions

$$\epsilon < \frac{N_c - 2}{N_c} \text{ and } \frac{d_{\min}^2}{\sigma_n^2} > -2\ln 2\left(1 - \frac{N_c\epsilon}{N_c - 2}\right), \tag{A.6}$$

$W(\boldsymbol{\alpha}|\mathbf{x}_0, \mathbf{y}) - W(\boldsymbol{\beta}|\mathbf{x}_0, \mathbf{y}) > 0$.

The conditions in (A.3), (A.4), (A.5), and (A.6) can be summarized and reduced to $\gamma > f^{-1}\left(\frac{1}{2N_c}\right)$ and $\frac{d_{\min}^2}{\sigma_n^2} > \max\left(\frac{2\sqrt{\ln N_c}}{\sqrt{2\gamma}-\sqrt{\ln N_c}}, 2\ln\frac{1}{1-2\epsilon}\right)$, where $f(x) = \frac{1}{x+2}\exp\{-\frac{x(x+1)}{x+2}\}$. Hence, Lemma 3.3 is proved.

References

1. S. Stein, "Unified analysis of certain coherent and noncoherent binary communications systems," *IEEE Transactions on Information Theory*, vol. 10, no. 1, pp. 43–51, Jan. 1964.
2. M. Simon and M. Alouini, "Exponential-type bounds on the generalized marcum q-function with application to error probability analysis over fading channels," *IEEE Transactions on Communications*, vol. 48, no. 3, pp. 359–366, Mar. 2000.

Appendix B
Calculation of the Probability Matrix W_1

Let $\mathbf{x} = \boldsymbol{\alpha} s$, $\mathbf{J} = \boldsymbol{\beta} b$ with $\boldsymbol{\alpha} = v(k)$, $\boldsymbol{\beta} = v(j)$, and $j, k \in \mathcal{I}_c$, $s, b \in \Omega$. Assume Ω is an M-PSK constellation with power P_s.

(i) When $j = k$, the received signal in the ith channel, r_i, and the corresponding Z_i defined in (3.80) can be calculated as

$$r_i = \begin{cases} s + b + n_k, & i = k, \\ n_i, & i \neq k, \end{cases} \quad Z_i = \begin{cases} \frac{\|b+n_k\|}{\sqrt{\|s+b\|^2+\sigma_n^2}}, & i = k, \\ \frac{\|n_i - s\|}{\sigma_n}, & i \neq k. \end{cases} \tag{B.1}$$

Note that $n_1, \ldots, n_{N_c}$ are i.i.d. circularly symmetric Gaussian random variables of zero mean and variance σ_n^2. For any $s, b \in \Omega$, Z_k is a Rician random variable with PDF $p_{Z_k}(z_k) = \frac{z_k}{\sigma^2} e^{-\frac{z_k^2+\nu^2}{2\sigma^2}} I_0\left(\frac{z_k \nu}{\sigma^2}\right)$ for $z_k \geq 0$, where $\nu = \frac{\sqrt{P_s}}{\sqrt{\|s+b\|^2+\sigma_n^2}}$ and $\sigma = \frac{\sigma_n}{\sqrt{2(\|s+b\|^2+\sigma_n^2)}}$; for $i \neq k$, Z_i's are i.i.d. Rician random variables with PDF $p_{Z_i}(z_i) = \frac{z_i}{\sigma^2} e^{-\frac{z_i^2+\nu^2}{2\sigma^2}} I_0\left(\frac{z_i \nu}{\sigma^2}\right)$ for $z_i \geq 0$, where $\nu = \frac{\sqrt{P_s}}{\sigma_n}$ and $\sigma = \frac{1}{\sqrt{2}}$. We have

$$\begin{aligned} W_1(k|k,k) &= \frac{1}{|\Omega|^2} \sum_{s\in\Omega} \sum_{b\in\Omega} Pr\{Z_k < Z_i, \forall i \in \mathcal{I}_c, i \neq k | \mathbf{x} = \boldsymbol{\alpha} s, \mathbf{J} = \boldsymbol{\beta} b\} \\ &= \frac{1}{|\Omega|^2} \sum_{s\in\Omega} \sum_{b\in\Omega} \int_0^\infty \left[Q_1\left(\frac{\sqrt{2P_s}}{\sigma_n}, \sqrt{2} z_k\right) \right]^{N_c-1} \frac{2(\|s+b\|^2+\sigma_n^2) z_k}{\sigma_n^2} \\ &\quad \cdot e^{-\frac{(\|s+b\|^2+\sigma_n^2) z_k^2 + P_s}{\sigma_n^2}} I_0\left(\frac{2 z_k}{\sigma_n^2} \sqrt{P_s(\|s+b\|^2+\sigma_n^2)}\right) dz_k, \end{aligned} \tag{B.2}$$

T. Li et al., *Wireless Communications Under Hostile Jamming: Security and Efficiency*, https://doi.org/10.1007/978-981-13-0821-5

where Q_1 is the Marcum Q-function and I_0 is the modified Bessel function of the first kind with order zero. For M-PSK constellation with power P_s, we have $s = \sqrt{P_s}e^{j\frac{2\pi m_s}{M}}$ and $b = \sqrt{P_s}e^{j\frac{2\pi m_J}{M}}$ where $m_s, m_J \in [0, M-1]$, then (B.2) can be simplified as

$$W_1(k|k,k) = \frac{1}{M}\sum_{\kappa=0}^{M-1}\int_0^{\infty}\left[Q_1\left(\frac{\sqrt{2P_s}}{\sigma_n}, \sqrt{2}z_k\right)\right]^{N_c-1}\frac{2[2P_s(1+\cos\frac{2\pi\kappa}{M})+\sigma_n^2]z_k}{\sigma_n^2}$$
$$\cdot e^{-\frac{[2P_s(1+\cos\frac{2\pi\kappa}{M})+\sigma_n^2]z_k^2+P_s}{\sigma_n^2}} I_0\left(\frac{2z_k}{\sigma_n^2}\sqrt{P_s\left[2P_s\left(1+\cos\frac{2\pi\kappa}{M}\right)+\sigma_n^2\right]}\right)dz_k, \tag{B.3}$$

where $\kappa \triangleq (m_s - m_J) \mod M$ is uniformly distributed over $[0, M-1]$. Since Z_i's are i.i.d. $\forall i \in \mathcal{I}_c, i \neq k$, then

$$W_1(i|k,k) = \frac{1}{N_c-1}[1 - W_1(k|k,k)], \ \ \forall i \in \mathcal{I}_c, i \neq k. \tag{B.4}$$

Therefore, we have **(P1)**: $W_1(k|k,k)$ and $W_1(i|k,k)$ are fixed values for any $i, k \in \mathcal{I}_c, i \neq k$.

(ii) When $j \neq k$, the received signal r_i and corresponding Z_i can be calculated as

$$r_i = \begin{cases} s + n_k, \ i = k, \\ b + n_j, \ i = j, \\ n_i, \quad\quad i \neq j, k, \end{cases} \quad Z_i = \begin{cases} \frac{\|n_k\|}{\sqrt{P_s+\sigma_n^2}}, \ \ i = k, \\ \frac{\|b-s+n_j\|}{\sqrt{P_s+\sigma_n^2}}, \ i = j, \\ \frac{\|n_i-s\|}{\sigma_n}, \quad i \neq j, k. \end{cases} \tag{B.5}$$

For any $s, b \in \Omega$, Z_k is a Rayleigh random variable with PDF $p_{Z_k}(z_k) = \frac{z_k}{\sigma^2}e^{-\frac{z_k^2}{2\sigma^2}}$, where $\sigma = \frac{\sigma_n}{\sqrt{2(P_s+\sigma_n^2)}}$; Z_j is a Rician random variable with PDF $p_{Z_j}(z_j) = \frac{z_j}{\sigma^2}e^{-\frac{z_j^2+\nu^2}{2\sigma^2}} I_0\left(\frac{z_j\nu}{\sigma^2}\right)$, where $\nu = \frac{\|b-s\|}{\sqrt{P_s+\sigma_n^2}}$ and $\sigma = \frac{\sigma_n}{\sqrt{2(P_s+\sigma_n^2)}}$; for $i \neq j, k$, Z_i's are i.i.d. Rician random variables with PDF $p_{Z_i}(z_i) = \frac{z_i}{\sigma^2}e^{-\frac{z_i^2+\nu^2}{2\sigma^2}} I_0\left(\frac{z_i\nu}{\sigma^2}\right)$, where $\nu = \frac{\sqrt{P_s}}{\sigma_n}$ and $\sigma = \frac{1}{\sqrt{2}}$. Then, $W_1(k|k,j)$ can be calculated as

$$W_1(k|k,j) = \frac{1}{|\Omega|^2}\sum_{s\in\Omega}\sum_{b\in\Omega} Pr\{Z_k < Z_j \text{ and } Z_k < Z_i, \forall i \in \mathcal{I}_c, i \neq k, j|\mathbf{x}, \mathbf{J}\}$$
$$= \frac{1}{M}\sum_{\kappa=0}^{M-1}\int_0^{\infty} Q_1\left(\frac{2}{\sigma_n}\sqrt{P_s\left(1-\cos\frac{2\pi\kappa}{M}\right)}, \frac{z_k\sqrt{2(P_s+\sigma_n^2)}}{\sigma_n}\right)$$

$$\cdot Q_1^{N_c-2}\left(\frac{\sqrt{2P_s}}{\sigma_n}, \sqrt{2}z_k\right)\frac{2z_k(P_s+\sigma_n^2)}{\sigma_n^2}e^{-\frac{(P_s+\sigma_n^2)z_k^2}{\sigma_n^2}}dz_k, \tag{B.6}$$

and

$$\begin{aligned} W_1(j|k,j) =& \frac{1}{|\Omega|^2}\sum_{s\in\Omega}\sum_{b\in\Omega} Pr\{Z_j < Z_k \text{ and } Z_j < Z_i, \forall i \in \mathcal{I}_c, i \neq j, k|\mathbf{x}, \mathbf{J}\} \\ =& \frac{1}{M}\sum_{\kappa=0}^{M-1}\int_0^{\infty} e^{-\frac{(P_s+\sigma_n^2)z_j^2}{\sigma_n^2}} Q_1^{N_c-2}\left(\frac{\sqrt{2P_s}}{\sigma_n}, \sqrt{2}z_j\right)\frac{2z_j(P_s+\sigma_n^2)}{\sigma_n^2} \\ &\cdot e^{-[\frac{P_s+\sigma_n^2}{\sigma_n^2}z_j^2+\frac{2P_s}{\sigma_n^2}(1-\cos\frac{2\pi\kappa}{M})]} I_0\left(\frac{2z_j}{\sigma_n^2}\sqrt{2P_s\left(1-\cos\frac{2\pi\kappa}{M}\right)(P_s+\sigma_n^2)}\right)dz_j. \end{aligned} \tag{B.7}$$

Since Z_i's are i.i.d. for any $i \in \mathcal{I}_c, i \neq j, k$, then

$$W_1(i|k,j) = \frac{1}{N_c-2}[1 - W_1(k|k,j) - W_1(j|k,j)], \quad i \in \mathcal{I}_c, i \neq j, k. \tag{B.8}$$

Therefore, we have **(P2)**: $W_1(k|k,j)$, $W_1(j|k,j)$ and $W_1(i|k,j)$ are fixed values for any $i, j, k \in \mathcal{I}_c$, $j \neq k, i \neq j, k$.

Appendix C
Subchannel Selection with Nonuniform Preferences

This appendix provides an approach to select K out of N_c subchannels according to a probability vector $\boldsymbol{\omega} = [\omega_1, \omega_2, \ldots, \omega_{N_c}]$, where ω_m denotes the probability that the mth subchannel is selected each time, and $\sum_{m=1}^{N_c} \omega_m = K$. Suppose ω_m's are rational numbers, then there exists a finite positive integer M, such that $l_m = M\omega_m$ is a positive integer for all $1 \leq m \leq N_c$. Furthermore, we have $\sum_{m=1}^{N_c} l_m = KM$. The approach works with the following steps:

1. Construct a $K \times M$ matrix, in which the kth $(1 \leq k \leq M)$ column represents the kth subchannel selection result; prepare l_m balls labeled "subchannel m" for all $1 \leq m \leq N_c$, and there are $\sum_{m=1}^{N_c} l_m = KM$ balls in total;
2. Initialization: set $k = 1$ as the current row to be filled, $m = 1$ as the current subchannel to be worked on, and $r = M$ as the number of empty entries for the current row;
3. Select l_1 entries randomly from the 1st $(k = 1)$ row of the matrix, and fill them with all the l_1 balls. For $k \geq 1$ and $m \geq 2$, placement of the l_m balls labeled "subchannel m" has two cases:
   - If $l_m \leq r$, the current row has a capacity large enough to accommodate all the l_m balls. Select l_m entries randomly from the kth row of the matrix, and fill them with all the l_m balls. Update the number of empty entries for the current row by $r \leftarrow (r - l_m)$; if all empty entries of the current row are filled, move to the next row by setting $k \leftarrow (k + 1)$ and $r \leftarrow M$.
   - If $l_m > r$, the l_m balls have to be split into the current row and the next row. First fill the r empty entries of the kth row with r out of l_m balls, then select $l_m - r$ out of $M - r$ entries randomly from the $(k + 1)$th row, and fill them with the remaining $l_m - r$ balls. Note that there are only $M - r$ entries in the new row available here, since the r columns already containing balls labeled "subchannel m" have to be avoided. Update the number of empty entries for the current row by $r \leftarrow [M - (l_m - r)]$, and set the current row by $k \leftarrow (k+1)$.

T. Li et al., *Wireless Communications Under Hostile Jamming: Security and Efficiency*, https://doi.org/10.1007/978-981-13-0821-5

4. Set $m \leftarrow (m+1)$ and repeat step 3 until all KM balls are placed in the $K \times M$ matrix;
5. Fetch each column in the matrix to generate the subchannel selection results for M consecutive time slots, and repeat all the steps above until all information transmission is done.

In the following, we justify that the probability of the mth subchannel being selected each time is exactly the desired ω_m. For each possible $1 \leq m \leq N_c$, the number of balls labeled "subchannel m" is $l_m = M\omega_m \leq M$. According to the approach above, all the l_m balls can be placed into at most two rows in the matrix. Denote $\mathcal{P}_{m,k}$ as the probability that the mth subchannel is chosen in the kth row. Then $\mathcal{P}_{m,k} = \frac{r_k}{M}$, where r_k is the number of balls labeled "subchannel m" that have been placed in the kth row of the matrix, since the mth subchannel would appear r_k times in the kth place out of the total M times of subchannel selection. If the l_m balls are placed into only one row, e.g., the k_0th row, for each subchannel selection, $\mathcal{P}_{m,k} = \frac{l_m}{M}$ for $k = k_0$, and zero elsewhere. Hence, the probability that the mth subchannel is selected considering all possible places would be $\mathcal{P}_m = \sum_{k=1}^{K} \mathcal{P}_{m,k} = \mathcal{P}_{m,k_0} = \frac{l_m}{M} = \omega_m$. If they are placed into two consecutive rows, e.g., the k_0th row and the (k_0+1)th row, then $\mathcal{P}_{m,k} = \frac{r}{M}$ for $k = k_0$, $\mathcal{P}_{m,k} = \frac{l_m - r}{M}$ for $k = k_0 + 1$, and zero elsewhere. In this case, $\mathcal{P}_m = \sum_{k=1}^{K} \mathcal{P}_{m,k} = \mathcal{P}_{m,k_0} + \mathcal{P}_{m,k_0+1} = \frac{r}{M} + \frac{l_m - r}{M} = \omega_m$. As a result, we can conclude that the probability that the mth subchannel is selected resulted from the approach is $\mathcal{P}_m = \omega_m$.

Example Suppose there are $N_c = 8$ subchannels, each time we select $K = 4$ out of $N_c = 8$ subchannels according to a subchannel selection probability vector $\boldsymbol{\omega}_s = [\omega_1, \omega_2, \ldots, \omega_8] = [0.9, 0.8.0.7, 0.6, 0.4, 0.3, 0.2, 0.1]$. In this case, $M = 10$, and we construct a 4×10 matrix. Furthermore, we prepare 40 balls, in which $M\omega_1 = 9$ balls are labeled "subchannel 1," $M\omega_2 = 8$ balls are labeled "subchannel 2," and so on. As illustrated in Fig. C.1, we first place all the nine balls labeled "subchannel 1" in nine entries randomly selected from the first row. Second, place one ball labeled "subchannel 2" in the remaining one entry of the first row, and the remaining seven balls labeled "subchannel 2" in seven entries randomly selected from the second row. Note that the columns already containing a ball labeled "subchannel 2" in the first row need to be avoided, and in the particular case with Fig. C.1 it is column 6. Repeat the procedure above until all the balls are properly placed in the matrix. Then, each column would indicate the selected subchannel indices

Fig. C.1 Example on subchannel selection with nonuniform preferences. (© [2016] IEEE. Reprinted, with permission, from Ref. [1])

	1	2	3	4	5	6	7	8	9	10
1	1	1	1	1	1	2	1	1	1	1
2	3	2	2	2	2	3	2	3	2	2
3	4	3	4	3	4	4	3	4	4	3
4	7	5	8	5	5	6	6	5	7	6

for one subchannel selection result. The entire matrix provides the subchannel selection results for ten time slots, and we can repeat all the steps above to generate more subchannel selection results. It can be verified that the probability for each subchannel being selected is exactly the one indicated in the subchannel selection probability vector.

Reference

1. T. Song, W. E. Stark, T. Li, and J. K. Tugnait, "Optimal multiband transmission under hostile jamming," *IEEE Transactions on Communications*, vol. 64, no. 9, pp. 4013–4027, Sept 2016.

Appendix D
Uniqueness of the Solution to Theorem 6.3

It is stated in Theorem 6.3 that: assuming there are K_s subchannels activated by the authorized user and K_J subchannels interfered by the jammer, the saddle point is reached when both authorized user and the jammer choose to apply uniform subchannel selection and uniform power allocation strategy. In this appendix, we show the uniqueness of this solution.

First, it is impossible to have nonuniform subchannel selection in a saddle point strategy pair. The reason is that: if either the authorized user or the jammer applies nonuniform subchannel selection, the nonuniform pattern could be detected and utilized by the other. That is, the authorized user could avoid the subchannels that are highly likely interfered by the jammer, while the jammer would prefer to interfere the subchannels that are highly likely used by the authorized user.

Second, with uniform subchannel selection for both the authorized user and the jammer, it is impossible to have nonuniform power allocation in a saddle point strategy pair. We will start with the case where the authorized user tries to maximize its capacity. We assume that the jamming power allocation (not necessarily uniform) is characterized by $\mathbf{P}_J^* = [P_{J,1}^*, P_{J,2}^*, \ldots, P_{J,K_J}^*]$, and different jamming power levels are assigned to the jammed subchannel randomly. That is, if a subchannel is jammed, the jamming power could be any $P_{J,m}^*(1 \leq m \leq K_J)$ with equal probability, $\frac{1}{K_J}$. The reason for random assignment is similar to uniform subchannel selection, i.e., the signal power allocation with a fixed pattern could also be detected and utilized by the jammer. Applying a similar idea to (6.24) and considering all possible jamming power for each jammed subchannel, the capacity of the authorized user can be calculated as

T. Li et al., *Wireless Communications Under Hostile Jamming: Security and Efficiency*, https://doi.org/10.1007/978-981-13-0821-5

$$
\begin{aligned}
C(\mathbf{P}_s, \mathbf{P}_J^*) = & \sum_{n=1}^{K_s} \left[\frac{K_J}{N_c} \sum_{m=1}^{K_J} \frac{1}{K_J} \frac{B}{N_c} \log_2 \left(1 + \frac{P_{s,n}}{P_{J,m}^* + P_N/N_c} \right) \right. \\
& \left. + \left(1 - \frac{K_J}{N_c} \right) \frac{B}{N_c} \log_2 \left(1 + \frac{P_{s,n}}{P_N/N_c} \right) \right] \\
= & \frac{1}{N_c} \frac{B}{N_c} \sum_{m=1}^{K_J} \sum_{n=1}^{K_s} \log_2 \left(1 + \frac{P_{s,n}}{P_{J,m}^* + P_N/N_c} \right) \\
& + \left(1 - \frac{K_J}{N_c} \right) \frac{B}{N_c} \sum_{n=1}^{K_s} \log_2 \left(1 + \frac{P_{s,n}}{P_N/N_c} \right).
\end{aligned}
\tag{D.1}
$$

Note that $\sum_{n=1}^{K_s} P_{s,n} = P_s$, and applying the concavity property proved in Lemma 6.1, we have

$$
\sum_{n=1}^{K_s} \log_2 \left(1 + \frac{P_{s,n}}{P_{J,m}^* + P_N/N_c} \right) \le K_s \log_2 \left(1 + \frac{P_s/K_s}{P_{J,m}^* + P_N/N_c} \right), \tag{D.2}
$$

and

$$
\sum_{n=1}^{K_s} \log_2 \left(1 + \frac{P_{s,n}}{P_N/N_c} \right) \le K_s \log_2 \left(1 + \frac{P_s/K_s}{P_N/N_c} \right). \tag{D.3}
$$

Substituting (D.2) and (D.3) into (D.1), we have

$$
\begin{aligned}
C(\mathbf{P}_s, \mathbf{P}_J^*) \le & \frac{K_s}{N_c} \frac{B}{N_c} \sum_{m=1}^{K_J} \log_2 \left(1 + \frac{P_s/K_s}{P_{J,m}^* + P_N/N_c} \right) \\
& + K_s \left(1 - \frac{K_J}{N_c} \right) \frac{B}{N_c} \log_2 \left(1 + \frac{P_s/K_s}{P_N/N_c} \right),
\end{aligned}
\tag{D.4}
$$

where the equality holds if and only if $P_{s,n} = \frac{P_s}{K_s}, \forall n$. So far, we have proved that: with random jamming power allocation, the authorized user can maximize its capacity only by uniform power allocation. For the jammer to minimize the capacity of the authorized user, we can obtain a similar result by applying the same method.

Appendix E
Proof of Lemma 6.3

To prove Lemma 6.3, we need the following result:

Lemma E.1 *For a real-valued function* $f(v) = \ln(1+v) - \frac{v}{1+v}$, $f(v) > 0$, *for any* $v > 0$.

Proof When $v > 0$, $f'(v) = \frac{v}{(1+v)^2} > 0$. Thus, $f(v) > f(0) = 0$. □

Now we are ready to prove Lemma 6.3.

1. The first-order derivative of $\widetilde{C}$ over K_s,

$$\frac{\partial \widetilde{C}}{\partial K_s} = \frac{K_J}{N_c}\frac{B}{N_c}\frac{1}{\ln 2}\left[\ln\left(1+\frac{\frac{P_s}{K_s}}{\frac{P_J}{K_J}+\frac{P_N}{N_c}}\right) - \frac{\frac{P_s}{K_s}}{\frac{P_s}{K_s}+\frac{P_J}{K_J}+\frac{P_N}{N_c}}\right] + \left(1-\frac{K_J}{N_c}\right)\frac{B}{N_c}\frac{1}{\ln 2}\left[\ln\left(1+\frac{\frac{P_s}{K_s}}{\frac{P_N}{N_c}}\right) - \frac{\frac{P_s}{K_s}}{\frac{P_s}{K_s}+\frac{P_N}{N_c}}\right]. \tag{E.1}$$

Let $v_1 = \frac{\frac{P_s}{K_s}}{\frac{P_J}{K_J}+\frac{P_N}{N_c}}$, then $\frac{v_1}{1+v_1} = \frac{\frac{P_s}{K_s}}{\frac{P_s}{K_s}+\frac{P_J}{K_J}+\frac{P_N}{N_c}}$. Similarly, let $v_2 = \frac{\frac{P_s}{K_s}}{\frac{P_N}{N_c}}$, then $\frac{v_2}{1+v_2} = \frac{\frac{P_s}{K_s}}{\frac{P_s}{K_s}+\frac{P_N}{N_c}}$. Applying Lemma E.1 to (E.1), we have

$$\frac{\partial \widetilde{C}}{\partial K_s} > 0, \text{ for any } K_s = 1, 2, \ldots, N_c. \tag{E.2}$$

2. The first-order derivative of $\widetilde{C}$ over K_J,

T. Li et al., *Wireless Communications Under Hostile Jamming: Security and Efficiency*, https://doi.org/10.1007/978-981-13-0821-5

$$
\begin{aligned}
\frac{\partial \widetilde{C}}{\partial K_J} =& \frac{K_s}{N_c}\frac{B}{N_c}\frac{1}{\ln 2}\left[\ln\left(1+\frac{\frac{P_s}{K_s}}{\frac{P_J}{K_J}+\frac{P_N}{N_c}}\right)-\ln\left(1+\frac{\frac{P_s}{K_s}}{\frac{P_N}{N_c}}\right)\right.\\
&\left.+\frac{\frac{P_s}{K_s}\frac{P_J}{K_J}}{\left(\frac{P_s}{K_s}+\frac{P_J}{K_J}+\frac{P_N}{N_c}\right)\left(\frac{P_J}{K_J}+\frac{P_N}{N_c}\right)}\right]\\
<& \frac{K_s}{N_c}\frac{B}{N_c}\frac{1}{\ln 2}\left[\frac{\frac{P_s}{K_s}\frac{P_J}{K_J}}{\frac{P_N}{N_c}\left(\frac{P_s}{K_s}+\frac{P_J}{K_J}+\frac{P_N}{N_c}\right)+\frac{P_s}{K_s}\frac{P_J}{K_J}}\right.\\
&\left.-\ln\left(1+\frac{\frac{P_s}{K_s}\frac{P_J}{K_J}}{\frac{P_N}{N_c}\left(\frac{P_s}{K_s}+\frac{P_J}{K_J}+\frac{P_N}{N_c}\right)}\right)\right].
\end{aligned}
\tag{E.3}
$$

Let $v_0 = \frac{\frac{P_s}{K_s}\frac{P_J}{K_J}}{\frac{P_N}{N_c}\left(\frac{P_s}{K_s}+\frac{P_J}{K_J}+\frac{P_N}{N_c}\right)}$, then $\frac{v_0}{1+v_0} = \frac{\frac{P_s}{K_s}\frac{P_J}{K_J}}{\frac{P_N}{N_c}\left(\frac{P_s}{K_s}+\frac{P_J}{K_J}+\frac{P_N}{N_c}\right)+\frac{P_s}{K_s}\frac{P_J}{K_J}}$. Applying Lemma E.1 to (E.3), we have

$$
\frac{\partial \widetilde{C}}{\partial K_J} < 0, \text{ for any } K_J = 1, 2, \ldots, N_c. \tag{E.4}
$$

Appendix F
Convergence Analysis of the Iterative Water Filling Algorithm

We prove the convergence of the iterative water filling algorithm (please refer to Table 6.1) by using the fact that an upper bounded and monotonically increasing sequence must converge. More specifically, we show that the constant c_2 in the algorithm is both upper bounded and monotonically increasing.

Since the total power of both the authorized user and the jammer is limited and the noise power levels are fixed, the water level, c_2, must be upper bounded. Next, we will show c_2 increases for each iteration. Recall that in the iterative water filling algorithm, c_2 is initialized with the solution obtained using the two-step water filling algorithm, and then we iteratively execute the following two steps:

$$\begin{cases} \frac{|H_{J,i}|^2}{|H_{s,i}|^2} P^*_{J,i} = \left[\min\left(\frac{|H_{J,i}|^2}{|H_{s,i}|^2} c_1, c_2 \right) - \sigma^2_{n,i} \right]^+, \ \forall i, & \text{(F.1a)} \\ P^*_{s,i} = \left(c_2 - \frac{|H_{J,i}|^2}{|H_{s,i}|^2} P^*_{J,i} - \sigma^2_{n,i} \right)^+, \quad \forall i. & \text{(F.1b)} \end{cases}$$

In the first iteration, we resolve violations against (6.46) by (F.1a). That is, for any subchannel with unnecessarily high jamming power (i.e., $\frac{|H_{J,i}|^2}{|H_{s,i}|^2} P^*_{J,i} + \sigma^2_{n,i} = \frac{|H_{J,i}|^2}{|H_{s,i}|^2} c_1 > c_2$), the unnecessary part $\frac{|H_{J,i}|^2}{|H_{s,i}|^2} c_1 - c_2$ will be moved to the less-filled subchannels (i.e., $\frac{|H_{J,i}|^2}{|H_{s,i}|^2} P^*_{J,i} + \sigma^2_{n,i} = \frac{|H_{J,i}|^2}{|H_{s,i}|^2} c_1 < c_2$). In this case, the total jamming power below the water level c_2 goes higher, which will inevitably *raise* the water level c_2 once (F.1b) is executed.

Starting from the second iteration, since c_2 has been increased by the previous iteration, when reallocating the jamming power by (F.1a), "jamming-efficient" subchannels will be "relaxed" (due to higher c_2) in the sense of being able to use more jamming power. A subchannel is said to be more jamming-efficient, if it has a lower jamming power path loss but higher user signal path loss, i.e., the ratio $\frac{|H_{J,i}|^2}{|H_{s,i}|^2}$

T. Li et al., *Wireless Communications Under Hostile Jamming: Security and Efficiency*, https://doi.org/10.1007/978-981-13-0821-5

is larger. For this reason, even the total original jamming power remains constant, the total "effective" jamming power below the water level c_2, which takes pass loss into account, would go higher again, due to increased contribution of highly jamming-efficient subchannels. This will again *raise* the water level c_2 once (F.1b) is executed. Now we are safe to say that the constant c_2 increases iteration by iteration.

Index

T. Li et al., *Wireless Communications Under Hostile Jamming: Security and Efficiency*, https://doi.org/10.1007/978-981-13-0821-5

Printed in the United States
By Bookmasters